中国 20 世纪建筑遗产项目 · 文化系列

China's 20th Century Architectural Heritage Project · Cultural Series

Centuried Keemun —"Tea" Stories of Cultural Chizhou

by CAH Editorial Department

悠远的祁红——文化池州的『茶』故事

《中国建筑文化遗产》编辑部 编

天津大学出版社
TIANJIN UNIVERSITY PRESS

在 60 多年前兴建的木质屋顶下，斑驳的铁皮机器如失去控制一般轰轰摇摆，混凝土柱子在一旁坚强地竖立，仿佛听惯了这嘈杂的“乐音”，随着皮带轮带动传送架缓缓运转，厚厚的茶渍仿佛在告诉人们这座活态遗产的价值。

目录 Contents

序一　Preface One

我是在2017年12月安徽池州召开的“第二批中国20世纪建筑遗产项目发布会”上，知晓池州“祁门红茶”老厂房的情况。今天看到即将付梓的《悠远的祁红——文化池州的“茶”故事》很感慨：一是中国文物学会20世纪建筑遗产委员会专家团队对“国润祁红”厂房的发现具有特别的意义；二是感慨池州市对文化遗产保护与发展的高度自觉与自信；三是在池州大地上不仅有40万辛勤耕种的茶农，更有守望1915年祁门红茶巴拿马万国博览会金质奖章的以殷天霁为代表的“国润祁红”企业员工的自豪感。

初读《悠远的祁红——文化池州的“茶”故事》一书，我明显感到有两个特点：其一，它为第一、二批中国20世纪建筑遗产的198个项目的传播带了好头。“国润祁红”20世纪50年代的旧厂房入选“中国20世纪建筑遗产项目”，说明其遗产珍贵、价值独特，具有代表性。其二，该书用“讲故事”的方式，阐述安徽的祁门是中国乃至世界红茶的发源地之一。珍视这个遗产是做强中国茶文化门类的历史和现实需要。中国20世纪工业建筑遗产是中国建筑的宝库，“国润祁红”旧厂房展示的既是传承的基础，又是文博艺术、文化旅游、工业遗产、历史人文诸方面创意的融合，需要政府的主导和文化政策的导向。

我相信该书是在文化传承与创意设计方面推动池州建设的重要力量，成为认知中国20世纪建筑遗产“国润祁红”的宣传指南，成为文化池州建设的一个示范读本。

单霁翔
中国文物学会会长
故宫博物院院长
2018年8月

I became aware of the condition of old factory of Keemun Black Tea in Chizhou when I participated the "Second Batch of China's 20th Century Architectural Heritage Conference" held in Chizhou. Seeing the forthcoming *Centuried Keemun—Tea Stories of Cultural Chizhou*, I feel rather emotional. Firstly, the discovery of Guorun Keemun factory by expert team from 20th Century Architectural Heritage Committee of Chinese Cultural Relics Society is of special significance. Secondly, I was impressed by high self-consciousness and self-confidence of Chizhou to protect and develop cultural heritage. Thirdly, besides 400,000 diligent tea farmers, Chizhou also owns pride from Guorun Keemun employees, who can be represented by Yin Tianji, for keep watching the gold medal of Keemun Black Tea in 1915 Panama International Exposition.

When I first read *Centuried Keemun—Tea Stories of Cultural Chizhou*, I obviously perceived two characteristics. Firstly, this book is a great start for the propaganda of first and second batch of 198 projects in China's 20th Century Architectural Heritage. As the old factory of Guorun Keemun in 1950s was selected into China's 20th Century Architectural Heritage Project, the preciousness and unique of the heritages are embodied. Secondly, this book interprets that Qimen, Anhui Province is one of the origins of black tea both in China and the world. Cherishing this heritage is the historical and practical need to strengthen Chinese tea culture. China's 20th century industrial architectural heritage is the treasure of Chinese architectures. What "Guorun Keemun" old factory shows to us is not only inheritance foundation but also an integration of innovative ideas of museology art, cultural tourism, industrial heritage and history humanities, which needs to be dominated by government and guided by cultural policies.

I believe that the book is an important driving force to construct Chizhou with cultural inheritance and innovative design. This book can become a propaganda guide for recognizing Guorun Keemun as China's 20th Century Architectural Heritage and become a model book of Chizhou cultural construction.

Shan Jixiang

President of Chinese Cultural Relics Society

Curator of the Palace Museum

August 2018

序二　Preface Two

自中国文物学会20世纪建筑遗产委员会发现“国润祁红”老厂房，到它入选“第二批中国20世纪建筑遗产名录”，再到我们在池州及北京故宫宝蕴楼进行研讨，“国润祁红”的历史科技与文化旅游价值愈发凸显。我想，池州山水画卷乃至茶韵之美，来自它节奏的慢、来自它生活的真、来自“国润祁红”厂房的特有风格与祁红人的文化坚守。今天《悠远的祁红——文化池州的“茶”故事》一书摆在面前，它应成为池州市向世人解读祁红茶“活态”文化的指南。

身为池州市人民政府的市长，我愈发感到，在这“千载诗人地”的沃土上有九华山佛教文化、有昭明太子封邑的士大夫文化、有好山好水的生态文化，而对于祁红乃至孕育它的贵池茶厂老厂房的遗产价值诠释得不足。从2018年元月中国文物学会20世纪建筑遗产委员会等单位编制的《文化池州建设——安徽国润茶业有限公司创意设计调研报告》中我看到，“国润祁红”之所以应成为令国人瞩目的“全遗产”，是因为它拥有一系列工业遗存与文化遗产方面的当代价值。“国润祁红”是池州应努力挖掘的最真实、最生动、最具特色的文化“点”，它保留至今，不仅是贵池厂的“宝”，更是池州市的“公共收藏”，体现了城市的精魂。

我相信读者读过此书，会喜欢祁红茶，会主动造访这清润怡人、悠远醇厚的传统文化之城，祁红的盛名更会与池州的文化发展之脉相契合。在此我感谢《悠远的祁红——文化池州的“茶”故事》创作团队为文化池州建设迈出的这一步。特此为序。

雍成瀚
安徽池州市委副书记、市政府市长
2018年8月

Since China's 20th Century Architectural Heritage Committee has found the treasure of Guorun Keemun old factory, the historical technology and cultural tourism value of Guorun Keemun have been more obvious, and then it was selected into the Second Batch of China's 20th Century Architectural Heritage Project and was researched in Chizhou and Baoyun Building in Beijing. From my perspective, Chizhou landscape paintings and even the beauty of tea rhythm derive from its slow pace, real life, unique features of Guorun Keemun old factory and cultural adherence of Keemun people. Today, the book *Centuried Keemun—Tea Stories of Cultural Chizhou* is right in front of me and this book should be a guide for citizens of Chizhou to interpret the live culture of Keemun Black Tea to the world.

Jiuhuashan Buddhist culture, culture of Zhaoming Prince's land distribution policy for scholar-official, ecological culture of wonderful landscape all boomed in the land of Chizhou. As the mayor of People's Government of Chizhou, I felt more and more that heritage value interpretation of old Guichi Tea Factory, which is the birthplace of Chinese Keemun, is far from enough. I clearly perceived the reason why "Guorun Keemun" becomes a spectacular national "Full Heritage" is the contemporary values from industrial and cultural heritage via the book *Construction of Cultural Chizhou—Research Report of Creative Design with Guorun Keemun Tea Co., Ltd. of Anhui* written by 20th Century Architectural Heritage Committee of Chinese Cultural Relics Society in Jan 2018. The live value of "Guorun Keemun" is the most authentic, vivid and unique cultural "point" that Chizhou should strive to explore. It being remained until today is not only the "collection" of Guichi factory, but also the "public collection" of Chizhou, which demonstrates the spirit of the city.

I believe that readers who finish reading this book will fall in love with Keemun Black Tea and visit this historical and pleasant cultural city. The reputation of Keemun will match with the cultural development of Chizhou, I would like to thank the creative team of *Centuried Keemun—Tea Stories of Cultural Chizhou* for making effort to develop cultural Chizhou. Hereby come with this preface.

Yong Chenghan
Vice Party Secretary and Mayor of Chizhou Municipality, Anhui Province
August 2018

前言 Foreword

历久弥新话祁红

Everlasting and Innovative Keemun

自1986年我于安徽农学院茶叶系毕业并任职安徽省贵池茶厂至今，不知不觉中32年过去了，正所谓“弹指一挥间”。

光阴荏苒，我这位20世纪80年代的新一辈已步入中年，很多事已记忆模糊了，但车间里溢满的祁红茶香，满负荷的工作，生产过程中各个工序、工位上的取样对比、讨论、调整，老茶师们神色宽厚地看茶做茶，乃至工余同事们结伴在秋浦河入长江口处中流击水，仍历历在目。

往事中个人沉浮事小，能亲历钟情一生的国润祁红茶业在大变革中把握时代脉搏，兴盛时保持清醒，遇到挫折之际不气馁、不放弃，终于迎来了稳健转身与健康发展，则是人一生的荣幸。

不可忘这家近百年茶号得到许多前辈的鼎力扶助。例如程家玉先生，中茶皖南分公司贵池茶厂时期的技术副厂长，1939至1948年间在祁门茶业改良场任技师；而在生产异常繁荣的情况下，唐克忠厂长仍不忘开发新产品的尝试，临时抽我到安庆师范学院联合开发祁红含气饮料，这是产业上极有远见之举。

我也曾离开茶厂从事茶叶管理工作，但在祁红厂遇到困境的时候，我决定回厂效力。那个时候几乎没人记起祁红是世界三大高香茶之一了，更没有为此打造的崇高目标。但作为一个学茶事茶的人，我回味车间里的祁红茶香，记得配制毛茶标准样时茶人们的敬业。我坚信：“这样的世界级好茶绝不能明珠暗投。”

祁红一直以出口高端市场为荣，但之后外贸的内耗竞争在一定程度上加剧了对行业的损害。其间红茶行业困境连连，同行中变故很多，比如有主打滇红的厂子适时地改为主打普洱，而贵池茶厂一直都在坚守祁红。

记得2007年长沙的中国国际茶业大会，在一片普洱、铁观音的汪洋中只有润思祁红一面红旗。福建茶叶协会会长詹立啖先生握着我的手说：“祁红还在的呀，你们真不容易。”胡平先生是主管过茶叶的商业部部长，他曾到我们公司审评，品了十来款茶，对我们那款“润思祁红·九五之尊”予以了致敬般的嘉许。

由于我们的坚守与创新，润思祁红入选了2010年“上海世博会中国世博十大名茶”，此为祁门红茶自1915年巴拿马博览会之后时隔95年再次闪耀世博会，为安徽茶赢得了荣誉，唤起了祁红的觉醒，安徽省农业产业化指导委员会专电祝贺。

2017年，我们代表徽茶参加外交部安徽全球推介活动。有人对此提出质疑，而主办方的回答则是：品质好，放心。于是就有了王毅部长的盛赞：“祁红——镶着金边的女王”的佳话。我们永远为祁红行业赋予正能量。

雍成瀚市长上任后不久，陪同中国摄影家协会主席李舸先生到公司，他率先指出老贵池茶厂应是池州的文化记忆。王宏书记专程到老厂区考察关心，之后，修龙、金磊等专家也在雍市长推荐下光临造访，又由此得到了文化大家单霁翔院长的重视喜爱。正是这些文化名流的参与，我们的主产品润思祁红系列不仅得到了业界的如潮好评，更使我们的老厂区入选“第二批中国20世纪建筑名录”。至此，国润茶业不仅是一种产业，更成为能推动城市产业发展的既传统又创新的文化。

日本茶学专家山西贞说祁红香是玫瑰香和木香的混合香，那仿佛就是森林里优雅的岁月味道，抑或是百年茶仓的馨香。

今值《悠远的祁红——文化池州的“茶”故事》一书出版，我代表本公司全体员工感谢此书编撰团队的辛勤劳作！同时，也不负各界厚望，继续前行在文化守望与创新之路上。

殷天霁

安徽国润茶业有限公司董事长

2018年7月

It has been 32 years with a fleeting instant since I graduated from the Department of Tea at Anhui Agricultural College and worked in Guichi Tea Factory in Anhui Province in 1986.

How time flies! I have now become a middle-aged man. Although lots of my memories became blurred, some things still appear vividly before my eyes, such as the fragrance of Keemun Black Tea in the workshop, full-load work, processes of produce chain, sampling, comparison, discussion and adjustment at the workstation, the kindly expression of the craftsmen when they were looking at and making tea, and even the memory of paddling at the entrance where Qiupu River flowed into Yangtze River.

Individual ups and downs doesn't matter, however, it is truly my pleasure to persist on my beloved life-long career—Keemun, which takes the pulse of the times in the great changes, remains clear-headed in prosperity and never gets discouraged or gives up in setbacks. Fortunately, Keemun has finally realized an amazing turnaround and realized healthy development.

We should never forget our century-old tea house was fortunate to get strong supports from many seniors within the industry, including Mr. Cheng Jiayu, Technical Deputy Director of Guichi Tea Factory of the Wannan Branch of China Tea Company. He worked as a technician in the improvement field of Keemun Tea Co,. Ltd. from 1939 to 1948. Even in the boom period of the factory, Tang Kezhong, the factory director, persisted in making attempts on developing new products. He transferred me to Anqing Normal College to jointly develop Keemun aerated beverage, which showed the foresight of the factory.

I once left the tea factory and worked elsewhere for tea management, but I decided to return to work at the tea factory when Keemun was confronted with difficulties. At that time, almost no one was aware that Keemun was one of the world's top three high-fragrant teas, let alone set a lofty goal for it. But as a person who is expert in tea, I has never forgot the fragrance of Keemun in the workshop and the dedication of workers when they made standard samples of primary tea. Therefore, I vowed to promote the world-class tea.

Keemun has always been proud of exporting to high-end markets, but the competitions in foreign trade have harmed the industry to some extent. As a result, quite a few tea factories got into trouble, for example, some factories that originally produced Yunnan Black Tea began to produce Pu'er Tea, while Guichi Tea Factory still persisted in Keemun.

I remember that Keemun Black Tea, the only black tea, was on displayed among Pu'er Tea and Tieguanyin at the China International Tea Industry Conference held in Changsha in 2007. Mr. Zhan Lidan, the President of the Fujian Tea Association, held my hand and said: "Keemun is still there, you are not easy." Mr. Hu Ping, an old expert who was Commerce Minister in charge of tea, once came to company to appraise a dozen of teas and spoke highly of our Imperial Throne of Runsi Keemun Black Tea.

Thanks to our persistence and product innovation, Runsi Keemun was selected as one of the top ten famous teas of Shanghai World Expo in 2010. Keemun Black Tea had once again shined in the World Expo since 1915 Panama Expo, which won honour for Anhui teas and aroused Keemun. Anhui Agricultural Industrialization Steering Committee sent a special congratulatory message to Keemun.

In 2017, we participated in the Anhui Global Promotion Activity organized by the Ministry of Foreign Affairs on behalf of Anhui teas. Although some people raised doubts about the qualification of Keemun for participating in the activity, the organizer believed that Keemun is a high-quality tea. Hence, China's Foreign Minister Wang Yi praised Keemun as Gilt-edged Queen. We will always inject positive energy to the Keemun industry.

Shortly after Yong Chenghan acted as mayor of Chizhou, he accompanied Mr. Li Ge, Chairman of the China Photographers Association, to visit the company and pointed out that the time-honoured Guichi Tea Factory should be the cultural memory of Chizhou. Also, Wang Hong, Secretary of the CPC Chizhou Municipal Committee, paid a special trip to the old factory. Recommended by Yong, other experts such as Xiu Long and Jin Lei were also invited to visit the factory. In addition, the factory has attracted attention from Shan Jixiang, Director of the Palace Museum. Due to the participation of these cultural celebrities, our main product Runsi Keemun Series have won the praise of the industry, and our old factory was selected into "Second Batch of China's 20th Century Architectural Heritage". At present, Guorun Tea Industrial Co., Ltd. represents not only an industry, but also a traditional and innovative culture that drives industrial development of the city.

A Japanese tea expert has ever said that the fragrance of Keemun was a mixture of rose aroma and wooden aroma, just like the elegant taste of times in forest, or the fragrance from a century-old warehouse of tea.

The book *Centuried Keemun — "Tea" Stories of Cultural Chizhou* is going to be published soon. On behalf of all the staff in the company, I would like to express my sincere thanks to the compiling team of this book for their hard work! Meanwhile, we will live up to the expectations of all and keep on going on the road of cultural inheritance and innovations.

Yin Tianji

President of Anhui Guorun Tea Industrial Co., Ltd.

July 2018

绪篇

“镶着金边的女王”产地的发现

在知识与信息可轻易获取的今日，树仍在生长，河水依然在流淌，云朵不时变换走向，人们却越来越希望回归生命本真的快乐。饮茶之风愈发高雅体面，更成为对自身心灵净化的精神之需。当茶的清香与醇厚甘甜隽永时，沁润在心田的每一寸便空灵舒静。本书将引领读者走进或许还鲜为人知的祁红世界，因为在外交部王毅部长赞美的“镶着金边的女王”中，人们还能品读到“祁红特绝群芳最，清誉高香不二门”的诗篇。应该说，池州是“国润祁红”的产地，祁红是自然赋予的宝物，重要的是我们该有所发现且在这里开启一次感悟茶史、品美味红茶的历程。

Continuation

Discovery of the Origin Place of "Gilt-edged Queen"

Knowledge and information are conveniently accessed to us nowadays, however, trees are still growing, rivers are still running, clouds are still changing directions. People are increasingly hoping to purse the real life being. Consuming tea becomes prominently elegant and decent and it has become a spiritual need for purifying soul. When the fragrance and mellowness of the tea are embraced, every corner of your heart will be tranquil. This book will lead readers into rarely-known world of Keemun since people can also experience, as described in the poem, "Preeminent Fragrance" of Keemun, which is complimented as "Gilt-edged Queen" by Foreign Minister, Wang Yi. Chizhou is the origin place of "Guorun Keemun" while Keemun is a treasure endowed by the nature. What more important is that we should make discover and start a journey where we can perceive tea history and taste delicious black tea.

远在与欧洲大陆尚隔一道海峡的英国，无论都市或乡间，每一处古老宅院似乎都保留着这样一番景象：一众绅士淑女高谈阔论，手中茶杯泛着红茶绚丽的汤色，飘扬着似花似果的芳香，而入口的回味更是醇厚隽永……一个多世纪前远销到欧洲的祁门红茶就是这样与莎士比亚、奥斯汀、雪莱、济慈等一道萦绕在人们的唇齿之间。

就我们自己民族沿袭数千年的社会生活而言，一则古老的民谚概括为“自古开门七件事,柴米油盐酱醋茶”，可知饮茶的确是中国各时代各阶层民众的生活必需品之一，帝王将相、士农工商、巫医百工、贩夫走卒……概莫能外。尤其值得注意的是，饮茶在中国不仅仅是生活之必需，更是一种文化生活的重要内容。种茶—饮茶—品茶，历代文人骚客每每将茶香、茶韵凝聚于笔墨毫端，时有精妙的品茶场景描绘，与中国式的诗情画意筋脉相连。诸多文学名著中，如“三言二拍”里写苏东坡与司马光斗茶、《红楼梦》里描绘妙玉烹茶的情景等，具有东方文化难以尽说的清雅韵味；近代作家老舍也在其话剧名作《茶馆》中尽写世间百态。茶叶无疑是我们民族文化意义上的象征物之一——礼仪之邦、诗之国度、丝绸之源、瓷器之鼻祖、茶叶之王国……

“中国”一词在拉丁文中写作“Chine”，按这个读音推测其汉字词源，至今众说纷纭——丝绸之国、瓷器之国、茶叶之国，不一而足。虽无定论，但丝绸、瓷器和茶叶这三项确实代表了西方人对中国的最初印象。

▨《陆羽烹茶图》，元代赵原绘

▨《惠山茶会图》，明代文徵明绘

▨《斗茶图》，元代赵孟頫绘

谁也不曾料到，在19世纪下半叶，茶叶作为中国本土芸芸众生之生活必需品和中国式清雅文化之象征，其盛其衰却已经不仅仅事关本国生计与文化消费，而是与西方列强的经济战略相关联，进而成为事关中华民族存亡的大事。那时候，西方列强正在由东西方正常的商业贸易转向以军事强权攫取在华经济利益，因为原本打算向中国倾销大量廉价工业产品以获利的英国人，却发

▨ 17 世纪，葡萄牙国王胡安四世的女儿，英国国王查理二世的王妃凯瑟琳·布拉干萨（1638–1705）作为将红茶传至英国的使者而留名青史。2016 年，葡萄牙分别以她及红茶为元素制作了五欧元硬币

现那时的中国对西方廉价工业品的需求量很有限，反倒是本国对中国茶叶的需求量却逐年攀升，由此造成了巨额贸易逆差。与亚洲相比，英国人对茶的第一印象不是人文美学的，而是物质文化的表象。[英] 艾伦·麦克法兰新近出版的《绿色黄金：茶叶帝国》，借由一个生活在印度阿萨姆茶产区的英国人类学家之笔，通过剖析英帝国的殖民统治，为英国将茶作为“奢侈品”和“武器”而赎罪。

这一现象早就引起了客居伦敦、正在研究西方社会命运的马克思的关注，他于 1853—1857 年多次投书《纽约每日评论报》。马克思在《对华贸易》一文中提到这样的史实：英国曼彻斯特机械化生产的布匹在中国没有市场，“世界上最先进的工厂制出的布匹的价格竟不能比最原始的织机上用手工织成的更便宜"。与此同时，马克思在《中国革命和欧洲革命》一文中陈述了这样的现象："从中国输入的茶叶在1793年还不超过16 167 331磅，然而在1845年便达到了50 714 657磅，1846年是57 584 561磅，现在（1853年）已超过6 000万磅。"其他资料显示，英国至1878年的茶叶消费量超过了1.36亿磅，一个收入中下等的普通英国家庭，平均每月的茶叶消费金额约占家庭总收入的10%。

正是这场不对等的国际商贸，令自命'先进''文明'

1657年，位于伦敦交易所巷（Exchange Alley）最先售卖茶叶的加韦咖啡屋（Garway's Coffee House）

的英国等列强采取很不文明的手段去扭转贸易逆差——对华倾销鸦片，并最终导致了东西方经济、政治、文化，直至军事等各个方面的全面冲突”。

就是在这样的历史背景下，英国等西方列强在施行倾销鸦片等不正当贸易行为的同时，也极力在中国境外寻求新的产茶基地以摆脱对中国茶叶的依赖，而中国茶业界则与其他行业一样，开始了一场历时一个半世纪以上的自我更新（1840 年至今），其中很值得大书特书的是祁门红茶。

在当今的茶业界，出产于安徽池州、徽州一带的祁门红茶堪称红茶中的极品，以其卓越的品质、迷人的香气，在国际上享有至高无上的声誉，与印度大吉岭红茶、斯里兰卡乌伐红茶并称为世界三大高香名茶。百余年来，祁红一直是英国王室的至爱饮品，其香名远播，每每被赞为“红茶皇后”，如诗人所言：“祁红特绝群芳最，清誉高香不二门。”

曾有中国营造学社耆老说：“喝滇红是为了记住抗战岁月，而品祁红则是对一种优雅文化的品味。”这个红茶世界的“群芳最”，有着一段跌宕起伏的“创业—守业—低谷—自我更新—再辉煌”的艰难路途，集中表现在以安徽池州为集中产地的“国润祁红”品牌中。

○ **别人喝茶，喝出和气，但鲁迅（1881–1936）喝茶，喝出怒气，享清福也成了讽刺。若剥去所有人眼中的光环下的鲁迅，他还是一个有抽烟、喝酒、饮茶喜好的人。**

有好茶喝，会喝好茶，是一种清福。不过要享这清福，首先必须有工夫，其次是练习出来的特别的感觉。

——《喝茶》

▨ **东方传来的红茶渐渐取代了酒精，成为英国皇室餐桌上的新时尚**

In each old house, regardless in city or countryside, of distant the United Kingdom, which is separated from the European continent by a channel, the following scenes seem to be remained: ladies and gentlemen are haranguing with a cup of black tea in hands. The tea glows with a gorgeous color, emits aroma of fruits or flowers, leaves a mellow aftertaste. The Keemun Black Tea, which was sold to Europe more than one century ago, lingered between people's lip and teeth along with Shakespeare, Austin, Shelley, Keats, etc.

▨ 1938 年 4 月 29 日，西南联大"湘黔滇旅行团"抵达昆明。次日，西南联大负责人与旅行团辅导团全体成员合影。一排左三为蒋梦麟，左五为梅贻琦，二排左五为闻一多

In terms of thousands years of social life of our own nation, it is summarized by an ancient proverb “Lives start with seven items, namely, firewood, rice, oil, salt, sauce, vinegar and tea”. Therefore, it can be observed that tea is indeed a must for all ranks of society, including emperors, generals, officials, farmers, workers, business men, wizards, doctors, peddler and all other people. In particular, tea is not only a must in daily life, but also an important part of cultural life. Plant tea-drink tea-taste tea, literati of various dynasties were

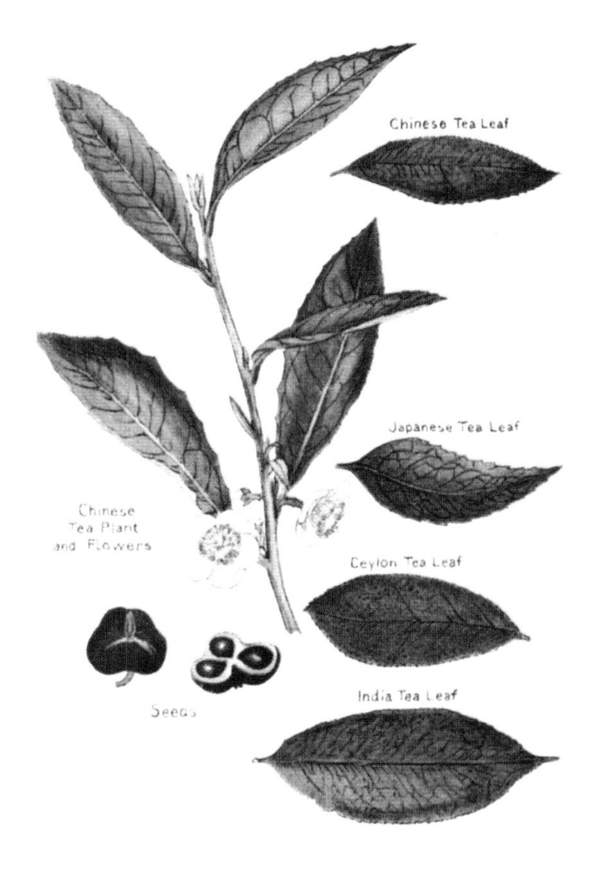

▨ 法国画家皮埃尔·约瑟夫·雷杜德（Pierre-joseph Redouté）绘制的中国茶及花朵图，他特别就中国茶叶及日本、斯里兰卡、印度等地产的茶叶的不同进行了对比。

fed by tea to create the stories about tea fragrance and tea rhythm under their pens. Lots of vivid tea scenes, blended with Chinese poetic illusion, were described in the literary works, for instance, the tea competition between Su Tungpo and Sima Guang in Sanyan Erpai and the tea-cooking scenario of Miao Yu in *Dream of Red Mansions*, which shows the unspeakable oriental elegance. Lao She, the modern writer, also described the world through famous drama *Tea House*. Tea is undoubtedly one of the significant symbols of our national culture—China is the kingdom of poetry, poem, silk, porcelain and tea.

The word "China" is written as "Chine" in Latin, and the source of its Chinese words is speculated according to this pronunciation. There have been lots of different opinions—the country of silk, the country of porcelain and the country of tea, etc. Although inconclusive, silk, porcelain and tea do represent the first impression of westerners on China.

Nobody had expected that the ups and downs of tea not only affected domestic livelihood and cultural consumption, but also was linked with economic strategy of western powers, thereby tea had become a critical matter for Chinese nation. In that era, the western powers were shifting from the normal commercial trade between the East and the West to grabbing economic interests in China by military power. British originally planned to dump a huge quantity of cheap

○ **徽州人胡适，在上海读书时所做的《藏晖室日记》，记录了他年轻时与朋友读书、喝茶、饮酒、看戏的琐事。他说他喜欢黄山毛峰、祁门红茶与太平猴魁等。**

胡适给族亲胡近仁的信中说："文人学者多嗜饮茶，可助文思。"这无疑是中国文人学者、高僧大德们的一个共识。徐志摩、郁达夫、林语堂等，都与胡适有交集。胡适除喜欢家乡徽州茶，还喜欢杭州龙井。程裕新是上海了不起的商号，20世纪末有人找到已发脆的1929年程裕新茶号编印的内部宣传册，胡适为其题有"恭祝程裕新茶号万岁"的字样，扉页还有孙中山先生对中国茶叶的简评。

industrial products to China, but found that Chinese demand for cheap industrial products was limited while British demand for Chinese tea increased yearly, which forming a huge trade deficit. Compared with Asian, the first impression of the British on tea is not humanistic aesthetics, but a superficial material representation. In the newly published book *Green Gold: The Empire of Tea* edited by [English] Alan Macfarlane, atonement was explained for UK taking tea as “luxury” and “weapons” through the life of a British anthropologist in the Assam tea producing area as well as the analysis of the British Empire’s colonial rule.

This phenomenon had long drawn the attention of Marx, who was living in London and studying the fate of western society and he repeatedly contributed to the *New York Daily Review* from 1853 to 1857. Marx mentioned such a historical fact in the article *Trade with China*: the fabrics produced by the way of mechanized production in Manchester, England have no market in China, “The fabrics price from the most advanced factory in the world even can’t be cheaper than the price of manual fabrics produced by the most primitive loom”. Meanwhile, Marx stated in the article *Chinese Revolution and European Revolution*: “The tea imported from China was less than 16,167,331 pounds in 1793, but in 1845 that number reached 50,714,657 pounds and 57,584,561 pounds in 1846. Now (1853) it has exceeded 60 million pounds.” According to

▨《绿色黄金：茶叶帝国》

Green Gold: The Empire of Tea

作者艾伦·麦克法兰(Alan Macfarlane)作为一位茶农兼茶商的遗孀，使得这本书具有一般研究著作所难以涉及之处。书中内容涵盖了关于茶叶以及茶叶贸易、茶叶传播和茶叶如何影响世界历史发展进程等有趣的内容。

▨ 崇山峻岭间一片宁静的茶园是祁门红茶的产地

materials, the tea consumption in the UK exceeded 136 million pounds by 1878. The monthly tea consumption of a lower-income British family accounted for 10% among the total household income on average.

It was this unequal international trade that made the British and other western powers, who praised themselves "advanced" and "civilized", adopt fairly uncivilized ways to reverse the trade deficit — dumping opium to China, which ultimately led to comprehensive conflict between the East and the West on economic, politic, culture and military.

In this historical context, the UK and other western powers were also trying to find new tea-producing bases outside China to get rid of the dependence on Chinese tea and meanwhile executed improper trade like dumping opium. Just like other industries, Chinese tea industry began a self-renewal of more than one and a half century (from 1840 to now),

▨ 加工过程中的祁门红茶

▨ 红茶被摆上餐桌

among which it was worthy of writing about Keemun Black Tea.

Keemun Black Tea, which is originated from Chizhou and Huizhou in Anhui province, is definitely preeminent among current tea industry. It enjoys a supreme reputation in the world for its excellent quality and charming aroma. Keemun Black Tea shares the title of Three High-fragrance Tea in the World with Indian Darjeeling Black Tea and Sri

▨ 浓郁的茶汤在容器中

▨ 身着汉服的女子在冲泡祁红茶

Lanka Wufa Black Tea. For more than a hundred years, Keemun has always been the favorite drink of the British royal family and its reputation has been far spread. Keemun is titled as "Queen of Black Tea", which is described as "The supremely fragrant tea" in the poem.

Mr. Qi from Chinese Construct Society said that "Tasting Dianhong Tea is to remember the days of Anti-Japanese War while tasting Keemun Black Tea is to enjoy an elegant culture." This "preeminent tea" has a dramatic history of "starting business—low point—self-renewal—glory again", which is demonstrated in "Guorun Keemun" based in Chizhou, Anhui province.

篇一

千载诗人地　百年红茶园
——双重遗产之贵池老茶厂

相对于文化遗产的修复与保存，最为困难的是对遗产的认定及坚忍不拔的传承，并使之“活在当下”。联合国教科文组织的《世界文化遗产地管理指南》中说：“每个世界遗产地都不止一个重要的故事来说明其历史：它们是如何被建造的或如何被破坏的，曾经生活在那里的人，曾经发生过的活动和事件……”国润祁红旧厂房的发现乃至逐步成为第二批中国20世纪建筑遗产项目，贵在它拥有真实感人的“活态”茶厂及品质祁红。老厂房虽不特意追求美学，但功能实用，具备了建筑本体的质朴之美。沿用至今的传统工艺与现代化生产线，形成了不需要雕琢堪称珍贵的“全遗产”样板。因为建筑遗产无形中都会诉说或歌唱，它颂扬守望着那些有“故事”的历史变迁与呵护者。

Passage One

Centuried Hometown of Poets and Black Tea
—Guichi Old Factory with Dual Heritage

Compared with the restoration and preservation of heritage, it is much more difficult to recognize, inherit with perseverance and make heritages alive nowadays. Management Guide for World Cultural Heritage Sites by United Nations Educational Scientific and Cultural Organization tells us "Each world heritage site has more than one important story to illustrate its history: how they were built or destroyed, the people once lived there and the activities and events happened there..."The valuable point of discovering Guorun Keemun old factory and being selected into Second Batch of China's 20th Century Architectural Heritage Projects is that it has a real touching live tea factory and high-quality Keemun . Although the old factory does not deliberately pursue aesthetics, it is in possession of unvarnished beauty of the architectures with its practical functions. The traditional manufacturing techniques, which are still applied today, are combined with modern production lines, forming a "full heritage" precious template which requires no modification. The architectural heritages will imperceptibly tell or eulogize the historical vicissitude and protectors with stories behind.

木梁下的高窗洒下唯一的一道光亮，麻袋中存放着毛茶垒砌如对局之初的棋子，墙壁上隐约看到工人们记录的数字，空气中弥漫着温润而浓郁的茶香，半世纪都未曾散去。

池州为安徽省地级市，别名“秋浦”，地处长江南岸之九华山地，下辖贵池区、东至县、石台县和青阳县，北与安庆市隔江相望，南接黄山市，西南与江西省九江市为邻。作为安徽省“两山一湖”（黄山、九华山、太平湖）旅游区的重要组成部分和中国佛教四大名山之一的九华山所在地，池州是长江下游地区重要的滨江港口城市之一。

池州素有“千载诗人地”之誉，至今留有大量的名胜古迹。自唐武德四年（621 年）正式设州置府，迄今近 1 400 年，而更早在南北朝时期，此地即为南梁昭明太子萧统（501—531 年）封邑，他曾住此编选《昭明文选》，这是中国现存最早的一部诗文总集；盛唐诗仙李白三上九华、五游秋浦，留下《秋浦歌》等众多诗篇；晚唐杜

第二首
晓起临妆略整容
提篮出户露方浓
小姑大妇同携手
问上松萝第几峰

第十九首
纵使愁肠似桔槔
且安贫苦莫辞劳
只图焙得新茶好
缕缕旗枪起白毫

第廿三首
乍暖乍凉屡变更
焙茶天色最难平
西山日落东山雨
道是多晴却少晴

第廿九首
茶品由来苦胜甜
个中滋味两般兼
不知却为谁甜苦
插破侬家玉指尖

○ “七山二水一分田”的徽州，男性早早便外出谋生，做着“徽骆驼”之工，采茶全由女性承担。从徽州茶叶的形态来考察，女性群体不仅承担了采茶工作，还受制于气候等限制。种植茶树的农家妇女还担负完成茶叶的初制加工。《采茶词》中有着如此形象描述。

▨ 由中国文物学会、中国建筑学会主办的"第二批中国20世纪建筑遗产项目发布"仪式于2017年12月2日在安徽省池州市召开，中国文物学会会长单霁翔、中国建筑学会理事长修龙等业内专家为入选项目代表颁牌。

牧曾任池州刺史，所作的《清明》使杏花村闻名于世。历代名人如陶渊明、苏轼、岳飞、陆游等都曾驻足池州，书写下珍贵的历史篇章。而贵池傩戏、"京剧鼻祖"青阳腔和东至花灯等一批国家级非物质文化遗产更名噪一时。不过，在上述与池州历史相关的物质与非物质文化遗产中，有一项或多或少是被世人所忽视的，即作为祁红主要种植、加工地之一的池州。但值得庆幸的是，这一缺憾于2017年8月后一步步得到了历史性的弥补。

2017年12月2日，由中国文物学会、中国建筑学会主办的"第二批中国20世纪建筑遗产项目发布暨池州生态文明研究院成立仪式"在安徽省池州市举办。中

由中国文物学会 20 世纪建筑遗产委员会与池州市人民政府联合举办的“文化池州工业遗产创意设计项目专家论证研讨会”一角

与会专家介绍茶厂建筑状况

国文物学会会长单霁翔、中国建筑学会理事长修龙等百位学界专家云集池州，在他们的见证下，位于池州的拥有 66 年历史的安徽国润茶业有限公司的祁门红茶老厂房——贵池茶厂，入选“第二批中国 20 世纪建筑遗产名录”。在此之前，安徽国润茶业有限公司的润思祁红制作技艺还成功入选“安徽省第五批省级非物质文化遗产代表性项目名录”。

在今长江南岸、池州市主城区（原贵池县）的西北一隅，坐落着一组具有新中国成立初期时代特色的老工厂建筑群。其毗邻池口码头的地理位置，令人不难推测出该老厂的产品大概是要经池口码头驶往长江航运线而发送外地的；厂区内呈锯齿形立面的老厂房，则为20世纪50年代苏式工业建筑的典型；距离锯齿形厂房不足一箭之地为工厂仓库，那里的每一间库房均是木板铺地、木质壁板环绕，观者至此无不被感染，每一块木板

○ **闻一多（1899–1946年）将喝茶视为生活中最重要的事。茶是生活的尺度，缺少茶的日子不知如何过。**

冰心在回忆中说，1930年夏，闻一多与梁秋实到我们燕京大学的新居来看望，我给他们倒上两杯凉水，闻一多忽然笑着说“我们出去一会儿”。当他们回来时，拿来一包茶，笑着说，以后要准备茶烟待客……闻一多为这个新成立的小家庭建立了一条烟茶待客的“风俗”。

与会嘉宾在同为“第二批20世纪建筑遗产”的故宫宝蕴楼前合影

▨ 拥有 66 年历史的安徽国润茶业有限公司的祁门红茶老厂房——贵池茶厂厂区入口

里都积存有浓郁的茶叶沁人心脾的馨香。至此，这座老厂房向世人揭示了它传奇般的身世：建于 1950 年的贵池茶厂，是属于农产品加工性质的制茶工厂。与此厂同时期同规模者，原本尚有祁门县祁门茶厂、东至县东至茶厂，但世事沧桑，时至今日而唯此独存。更珍贵的是，其前身就是同属池州地区范围的距离此地约 70 千米的原至德县尧渡街茶号——余干臣创始祁门红茶的源头所在。

▨ 尧渡老街及当年的红茶厂旧址

▨ 剖面呈锯齿形的制茶厂房

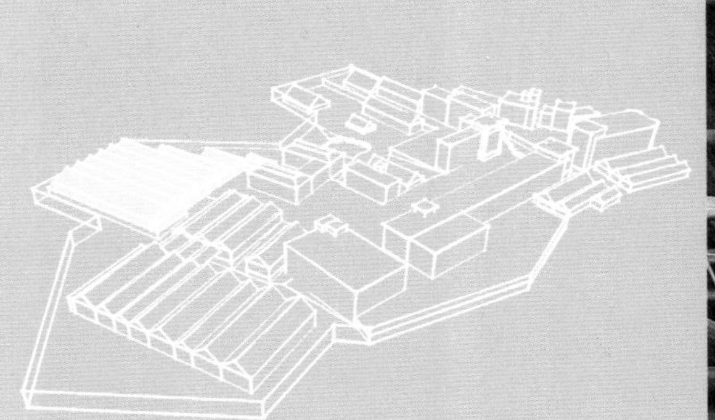

制茶车间

制茶车间是老厂区内占地面积最大的建筑，外观由六个一面坡屋顶并联组成大车间屋顶，立面呈锯齿状；室内由巨大的水泥柱形成无隔断墙的柱网空间。其一面坡顶为垂直的玻璃墙与斜坡之瓦面组成，水泥柱中空，实为排水管道。室内机械设备为1950年进口设备，其生产功能经专门设计，至今仍可使用。此车间保证了祁红生产“精制工艺”阶段的大规模生产量。就建筑本体而言，制茶车间明显受当年苏式工业建筑风格影响，外观简洁朴素，以保证实用功能为首要。

新机房

新机房1961年4月施工，1962年5月竣工。据老职工回忆当时有8个施工队同时施工，场面非常壮观。整体建筑规整，风格简约，较多地采用了新技术、新结构、新材料，代表了当时茶产业的最高水平。它的内部采用大跨度设计，以12根圆形廊柱为支撑。更为珍贵的是新机房内装备了当时国内最传统的祁门红茶联装生产流水线，这些制茶老设备得到了较好保护和适当使用。

New machinery room

The full name of the new machinery room is hand-picked field. Its history can be traced back to 1962.

The new machinery room was constructed from April 1961 to May 1962.The construction took 13 months. According to the elder workers' memory, the building was constructed by 8 teams at same time. The scene was very spectacular. Overall architectural style is neat, simple, using more new technologies, new structures, new materials, it is on behalf of the level of technological development of tea industry at that moment. Its interior design was used by large span and supported by 12 circular pillars. More valuable point is the new machinery room was equipped with the most traditional domestic Keemun-mounted production lines. These old tea making equipments have been well protected and properly used.

机别
电动机
规格型号
维修日期
备注

老毛茶仓库

老毛茶仓库建于1951年，为青砖墙体的库房建筑，由三个“人”字坡顶并联组成屋顶，外观正立面呈“山”字形，整体风格简洁明快。其室内相应为长廊居中，两侧分列12个尺度统一的内库房，其地板与四面壁板均为大兴安岭红松板材；长廊内现存大型过秤。那些与建筑同龄的红松板材，浸透并散发着让人难以忘怀的茶香。

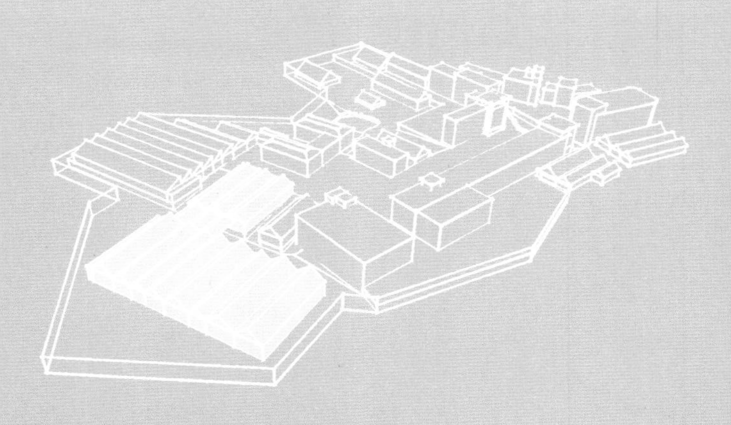

國潤

毛茶仓库

DB—3型
地中衡

國潤

生产重地

手工拣厂

手工拣厂是“精制工艺”阶段的生产场地，为二层楼建筑，下层室内矩形空间宽敞，四壁辟玻璃窗，采光明亮。居中一列略显细瘦的木柱上承横梁，而横梁由上下两层木板挟蜂巢状木格组成，具有简洁内敛且不失装饰性的美感。根据居中立柱分析，原设计为平房，后因生产规模扩大而加建上层，故为安全起见而增加立柱。

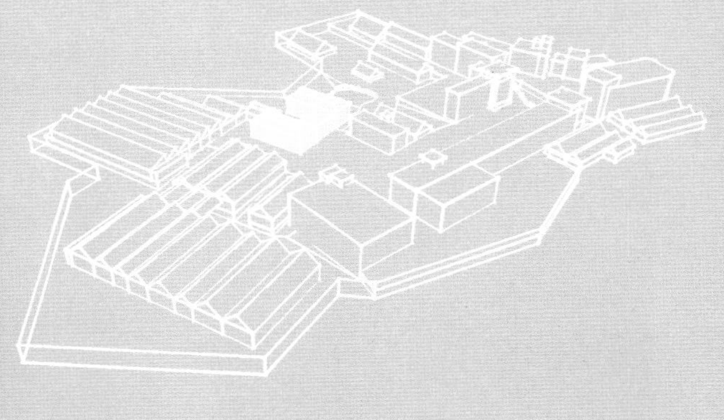

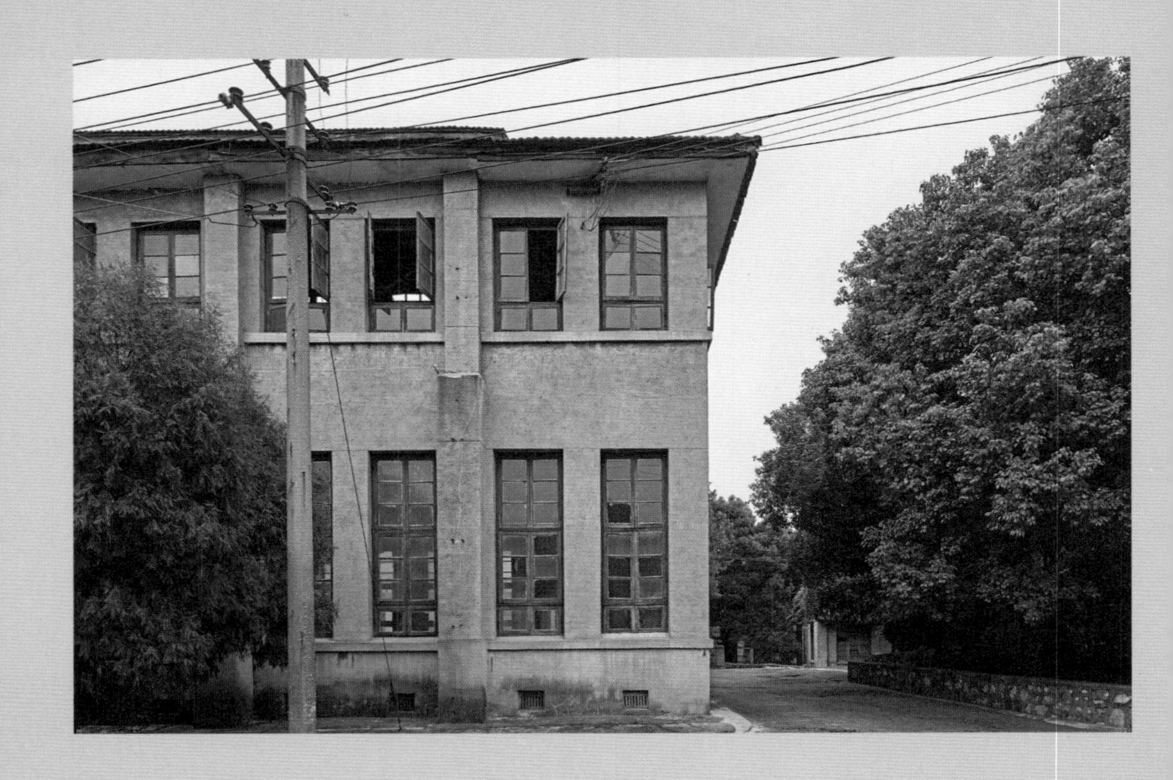

红茶发酵特征、程度比照(参考)

发酵特征	发酵程度
青绿色,有强烈青草气	开始发酵
青黄色,有青草气	发酵不足
黄色,略显清香	发酵偏嫩
黄红色,遗发花香或果香	发酵适度
红色,熟香	发酵偏老
暗红色,香气低沉	发酵过度

注:根据室内湿、温度、时间、发酵程度特征掌控

现存主要由制茶车间、老毛茶仓库、手工拣厂等组成的原贵池茶厂老厂区，建筑实体为1951年始建，是新中国成立初期引进欧洲、苏联等国外建筑技术所建造的工业建筑佳作，其简洁的外观与实用功能自然融合，为存世较少的现代工业建筑实例，客观记录了新中国成立初期的经济状况和外贸策略——农产品转向国营化生产，具有浓郁的现代工业建筑气息和现代建筑艺术风格，而其建筑功能却又是服务于古老的农产品加工的，寄托着几代人富国强民的理想。

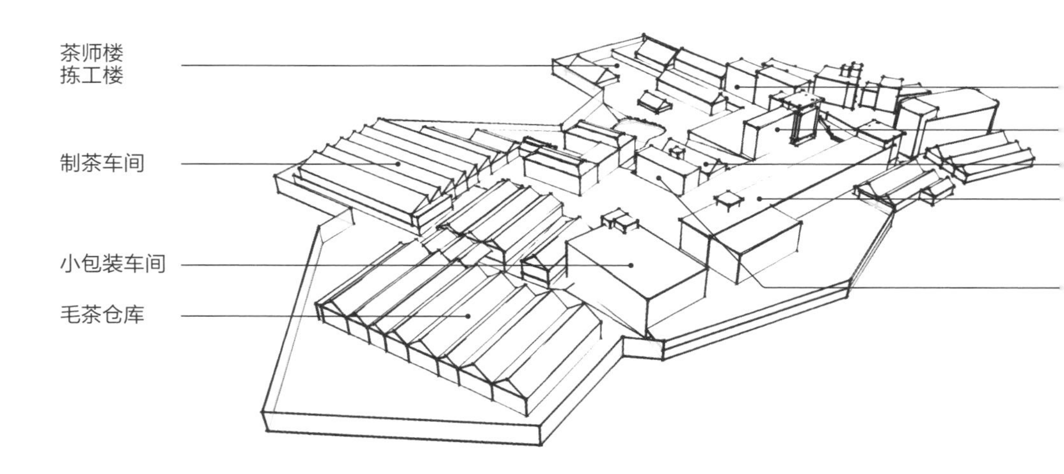

▨ 鸟瞰厂区布局

▨ 自出口仓库顶层俯视制茶车间，锯齿形的屋顶延绵不绝

▨ 茶师楼与拣工楼

—— 原职工宿舍

—— 办公楼

—— 职工之家

—— 制茶车间

—— 质检楼

▨ 时光的流逝，掩饰不住质检楼立面当年的精美

▨ 尽管当年的花式栏杆已经被青苔覆盖，琉璃一般的色彩依然昭示着往昔的兴盛

▨ 1953 年兴建的职工之家整体风格规整，立面简洁，内部采用全木质结构

▨ 办公楼采用外走廊的形式串联，栏杆花饰和立面选材都颇具时代特色

The prefecture-level city Chizhou in Anhui province, with an alias of Qiupu, lies in Jiuhuashan region on the southern bank of the Yangtze River. Chizhou has jurisdiction over Guichi district, Dongzhi county, Shitai county and Qingyang county. It faces Anqing city across the river to the north and Huangshan city to the south and it is adjacent to Jiujiang city of Jiangxi province in the southwest. As an important part of tourist area of "Two Mountains and One Lake" (Huangshan, Jiuhuashan, Taiping Lake) in Anhui and the locality of Jiuhuashan which is one of the Four Famous Buddhist Mountains in China, Chizhou is one of the important riverside-port cities in the lower reaches of the Yangtze River.

棕红色的屋顶灰、青色的石柱构成一幅和谐的画面，在一道后期添设的木梁下吊起的，是专为下方制茶员工准备的电扇

Chizhou is known as the "Land of Poets" and has lots of scenic spots and historical sites. Since the fourth year of Tang Wude (621), it was officially set up state and government and had a history of nearly 1,400 years so far. In earlier age of Southern and Northern Dynasties, Chizhou was a monarch-granted manor estate of Zhaoming (501–531), the prince of the Southern Liang. Chizhou was the place where Zhaoming complied *Zhaoming Album* which was the earliest existing set of poems in China. The poetic genius, Li Bai in the flourishing Tang Dynasty, climbed Jiuhuashan for three times and visited

夕阳西下，晚班的茶工拽一下灯绳，便可继续进行机器的操作

Qiupu for five times, leaving *Song of Qiupu* and many other poems; Du Mu in late Tang Dynasty wrote *Qingming* and made Xinghua village famous in the world. Celebrities, such as Tao Yuanming, Su Shi, Yue Fei and Lu You all had visited Chizhou and wrote precious historical articles. A number of national intangible cultural heritages, such as the Guichi Opera, the originator of Peking Opera—Qingyang Opera, Dongzhi Lanterns also gained considerable fame. However, one of above tangible and intangible cultural heritages related to the history of Chizhou is more or less overlooked by people, namely, Chizhou is also the main cultivated and processed land of Keemun. Fortunately, this pity has been historically compensated step by step after August 2017.

The "Ceremony of Second Batch of China's 20th Century Architectural Heritage Project and the Establishment of Chizhou Ecological Civilization Research Institute" hosted by the Chinese Cultural Relics Society and the Chinese Architectural Society was held in Chizhou, Anhui Province on December 2nd, 2017. Shan Jixiang, the president of Chinese Cultural Relics Society and Xiu Long, the director of Chinese Architectural Society along with other hundreds of academic experts gathered in Chizhou. With their witness, Guichi Tea

拣厂采用了精致的双层木屋梁结构，中间用拼接好的木条斜撑，再用长螺栓进行固定。纤细的木柱亦巧妙地偏向一侧而非居中，尽显结构之轻盈

▨ 机器的底座上留有“安全为革命，革命促安全”字样，显示着长此以往的坚守

Factory, the old tea factory of Keemun Black Tea possessed by the 66-year-old Guorun Tea Co., Ltd. of Anhui in Chizhou, were selected into Second Batch of China's 20th Century Architectural Heritage. Prior to this, manufacturing crafts of Runsi Keemun of Guorun Tea Co., Ltd. of Anhui were selected into the Fifth Batch of Provincial Intangible Cultural Heritage Representative Projects of Anhui Province.

A group of old factories with the characteristics of the early period of the PRC are located on the southern bank of the Yangtze River and the northwest of Chizhou (Former Guichi County). It could be speculated that the products of the old factories were probably sent to the Yangtze River and shipped to other places via Chikou Port according to its adjacent location to the Chikou Port. Old building with zigzag wall in the factory is typical Soviet-style industrial buildings in 1950s. Factory warehouses stand in a distance of a stone's throw from the zigzag factory buildings, each warehouse is paved with wooden planks and surrounded by wooden siding.

Visitors can be easily moved by refreshing aroma stored in each wooden plank. So far, the old factory reveals its legendary experience to the world: the Guichi Tea Factory, built in 1950, is a tea factory where agricultural products are processed. There were some contemporary factory with the same scale, such as Keemun Tea Factory in Keemun county, Dongzhi Tea Factory in Dongzhi county, but only Guichi Tea Factory is remained throughout historical vicissitudes. What is more

分筛机一侧悬挂着十余片不同粗细的筛网，茶工们踩上木质的二层小台阶便可方便地对筛网进行更换

valuable is that its predecessor, Yaodujie Teahouse where Yu Ganchen created Keemun Black Tea, was only 70 kilometers from here and belonged to Chizhou as well.

The existing old Guichi Tea Factory is mainly composed of tea making workshop, old Maocha warehouse, manual picking field, etc. The original building was built in 1951. It was a masterpiece of industrial building with the foreign construction technology from Europe and the Soviet Union. The Guichi Tea Factory with natural integration of succinct appearance and practical functions is a rare case of modern industrial architecture, objectively recording the economic situation and foreign trade strategy of the early PRC—agricultural products turned to state-owned production. The factory, with a strong sense of modern industrial building and a new architectural artistic style, serves for the processing of ancient agricultural products by its architectural function, entrusted with the ideals of founding a wealthy and powerful country by generations.

篇二

烽烟国门下　茶香化干戈
——祁门红茶的前世今生

1559年，威尼斯出版了一本波斯人写的《航海与旅行》："Chai catai（中国茶），它们有的是干的，有的是新鲜的，放在水中煮透……这种东西在喝的时候越烫越好。"这可能是西方对茶业最早的记载，但可能还不是西方人最早接受饮茶习俗的记录。此记录中，所谓"干的、新鲜的"，可能分指红茶、绿茶。至17世纪下半叶，茶叶不仅为毗邻中国的俄罗斯所接受，更经荷兰输入英国王室，迅速风靡欧洲（茶叶最初于1606年由荷兰引进欧洲大陆，1662年进入英国）。至此，茶叶逐渐与可可、咖啡并列，以独特品质参与其中的祁门红茶就开始登上舞台。

Passage Two

Turn Swords into Ploughshares with the Fragrance of Tea —Past and Present of Keemun Black Tea

*In 1559, Venice published a book **Navigation and Travel** written by Persians and it was said in this book: "Some of Chai catai (Chinese tea) are dry, some are fresh, and the tea is boiled in water... it is better to drink this kind of thing when it is hot." This might be the earliest record of the tea industry by the West, but it might not be the earliest record about westerners' acceptance for the custom of drinking tea. In this record, the so-called "dry" and "fresh" may refer to black tea and green tea separately. In the second half of the 17th century, tea was not only accepted by Russia, which was adjacent to China, but also was imported into the British royal family by the Netherlands and it quickly prevailed in Europe (tea was firstly introduced to the European continent by the Netherlands in 1606 and entered the United Kingdom in 1662). From this point, tea was gradually juxtaposed with cocoa and coffee. The Keemun Black Tea entered the stage with its unique quality.*

茶与茶文化

保守估计，中国人饮茶的历史至今已有3 000余年。至迟在公元780年唐代陆羽所著《茶经》问世之后，饮茶已成为中国特有的生活习惯和文化现象，并逐渐影响至全球。《茶经》不仅记载了唐代之前的中国茶叶生产、饮用的经验，更倡导、归纳出一种“精行俭德”的茶道精神，使得茶与中国书画等并列成为中国文化的象征。继《茶经》之后，北宋蔡襄（1012—1067）于北宋皇祐年间（1049—1053）著《茶录》，全书分上下篇论及色、香、味，藏茶、炙茶、碾茶、罗茶、候汤、熁盏、点茶等烹饮方法，以及茶焙、茶笼、砧椎、茶铃、茶碾、茶罗、茶盏、茶匙、汤瓶等器皿内容，是《茶经》之后最有影

公元780年左右，世界上最早的一部茶叶专著《茶经》问世

这是当时中国关于茶的经验总结。作者陆羽（733—804）详细收集历代茶叶史料、记述亲身调查和实践的经验，撰《茶经》三卷，对唐代及唐代以前的茶叶历史、产地、茶的功效、栽培、采制、煎煮、饮用的知识技术都作了阐述，是中国古代最完备的一部茶书。

《茶录》作于北宋皇祐年间（1049—1053）

《茶录》一书为作者“承陛下知鉴”所作，分上下两篇，其论述对后世影响深远。如在论述茶色时，他说：“茶色贵白，而饼茶大抵于表涂膏泽，故有青、黄、紫、黑之异。”论述茶香时，他说：“茶有真香，而入贡者，微以龙脑和膏以助香，建安民间试茶，皆不入香，恐夺其真。”

响的论茶专著。

虽然中国人如今的饮茶方式以沏泡清茶（绿茶、花茶）为主，但从全球范围来看，千差万别的饮茶方式及其所蕴含的不同的文化追求，都可在原产地中国找到源头。如当今的中国茶文化以品赏清淡幽香的绿茶为主，基本方式为热水冲泡，但在历史上，追求浓郁口感的煮茶饮用法。中国式的茶文化波及朝鲜、越南、日本等周边国家，后传入俄罗斯、印度、英国。当今世界，饮茶成为社会习俗并成为文化现象者，大致可按国家命名，俄罗斯、英国、德国、美国、土耳其、阿根廷等可称为重要的茶文化分支。

1. 俄罗斯茶文化

俄罗斯是最早从中国传入茶叶的国家（据记载，俄

俄罗斯人用“茶炊”来享用下午茶

“茶炊”是俄罗斯独特的文化和艺术标志。这种茶具不需要花太多的时间来煮茶，其中的水也能够保持温度，使茶味香浓。“茶炊”上有时摆放小茶壶，周边一般会放有俄式点心和果酱。

在英格兰宫廷中饮用红茶的凯瑟琳

罗斯人首次接触茶约在1638年）。俄罗斯人饮茶与中国区别较大，他们变中国本土流行的沏泡茶为煮茶，并发明了一种煮茶的工具——“茶炊”。而煮出来的茶虽浓郁，但难免苦涩，于是又在热茶汤中添加糖、柠檬汁乃至牛奶等。

2. 英国茶文化

英国略晚于俄罗斯，从17世纪60年代开始进口茶叶，到18世纪50年代，茶叶已经变成英国人的全民饮料，并由此派生出独特的英国“下午茶文化”。相传葡萄牙公主凯瑟琳（1638—1705）嗜饮中国红茶，于1662年嫁与英国国王查理二世，她的嫁妆中包括221磅产自中国福建的红茶及各种精美的中国茶具，而在那个时代，红茶之贵重堪比银子。英国诗人埃德蒙·沃尔特曾作一首赞美诗献给这位“饮茶王后”：

维纳斯的香桃木和太阳神的月桂树，
都无法与王后称颂的茶叶媲美。
我们由衷感谢那个勇敢的民族，
因为他们给予了我们一位尊贵的王后
和一种最美妙的仙草，
并为我们指点通往繁荣之途。

有皇室引领风尚，英国成为饮茶大国似乎就顺理成章了。如果说咖啡催生了法国的文艺与哲学思辨，则英式午茶是其诗歌、小说的灵感策源——有拜伦、济慈之诗意盎然，有奥斯汀式的机智与优雅，也有狄更斯笔下的贫街陋巷……

3. 土耳其茶文化

土耳其自产的茶叶属于红茶的一种。土耳其人与美国人相似，也是茶叶与咖啡并重，是伊斯兰教国家中的茶叶消费大国。土耳其人热情好客，请喝茶更是他们的一种传统习俗，也可以说，土耳其的茶文化展示着伊斯兰教国度热情好客的文化特色。

红茶与红茶文化

按照茶叶的成品分类，茶文化的世界形成了两大门派：以中国茶文化为代表的绿茶文化——延绵数千年的耕读传统与东方诗意性的“品茗”；以红茶（含半红茶）为主的英国、俄罗斯茶文化——适应高寒地带的浓郁与工业革命之后人们对优雅文化的追求。

1. 以绿茶为主的中国茶文化

中国并不纯饮绿茶，饮用衍生品花茶的消费数额也未必少于绿茶，而半发酵性质的半红茶（以福建乌龙茶、云南普洱茶等为代表）与全发酵性质的红茶（以福建武夷山岩茶为代表）也甚为流行，更有内蒙古、西藏、新疆等地的牧区主要消费半发酵的茶砖（奶茶、青稞茶等）。即使在纯粹绿茶的世界里，又有西湖龙井、碧螺春、黄山毛峰、六安瓜片、君山银针等有着微妙差异的诸多品种。但总体说，欣赏绿茶的纯天然味，从中体会大自然的恩赐与人世百态，是中国茶文化的根本。

2. 以红茶为主的英国、俄罗斯茶文化

同样，英国、俄罗斯等国家以饮用红茶为主，但也

▨ 英国19世纪初诗人拜伦肖像及中国演唱者在歌剧舞台上所扮演的唐璜形象

并不排斥其他。不过，其根本的着眼点，在于高寒地带所需要的浓郁，以及物质生活必需之余由品茶而引发的文化生活。俄罗斯大诗人普希金名作《叶甫盖尼·奥涅金》中有这样的诗句：

天色转黑，晚茶的茶炊
闪闪发亮，在桌上咝咝响，
它烫着瓷壶里的茶水；
薄薄的水雾在四周荡漾。
这时已经从奥尔加的手下
斟出了一杯又一杯的香茶，
浓酽的茶叶在不停地流淌
……

从中国舶来的茶在俄罗斯保持了本质的清香，但异域的茶炊却展示着与本土茶饮相比微妙别样的茶文化氛围。之后的托尔斯泰、契诃夫、高尔基以及音乐家柴可夫斯基、画家列宾等，无一不从这种文化氛围中孕育其传世之作。

英国不同于俄罗斯，是茶文化的另一个重要分支——英式午茶所派生出来的奥斯汀式的美文与拜伦、济慈、雪莱的诗歌。拜伦（George Gordon Byron，1788—1824）在其代表作《唐璜》中写道：

我觉得我的心儿变得那么富于同情，
我一定要去求助于武夷的红茶（Black Bohea）；
真可惜，酒却是那么的有害……

诗中提到的“武夷的红茶”，系指晚清时期中国专为应付外销市场而生产的专供出口的红茶。中国茶为这两个文化大国做出了贡献。

○ **对于中国红茶的外销，西方人在他们的书籍中也有所记录。**

据尤克斯《茶叶全书》中的“茶叶年表”所记述，1705年，爱丁堡金匠刊登广告，绿茶（Green Tea）每磅售十六先令，红茶（Black Tea）三十先令。英国传记作家玛丽返蓝尼夫人记录，当时茶价为红茶二十至三十先令。由于中国红茶茶味浓郁、独特，在国际市场上备受欢迎，远销英国、荷兰、法国等地。英国人诺顿夸奖说：“喝这种茶胜过饮人参汤。”

中国外销红茶的诞生

1. 西方人对茶的引进与培植

俄罗斯、英国等欧洲国家所饮之茶，最初由中国而来，历史可上溯至唐代，但真正风靡欧洲则在明代之后，距今约四百年左右，并形成欧洲以俄罗斯为代表的茶厨煮茶方式和英国的"晨茶·午茶·晚茶"饮茶习惯。初期使用的茶叶，基本上是半发酵的中国茶砖和全发酵的武夷山红茶。

2. 中国纯外销型红茶的诞生

中国在明代开始产生红茶，说起来倒是有个意外。相传明朝中后期的某年，地处武夷山区的福建省崇安县桐木地区在采茶的季节，当天已采摘的茶青因故没有来得及制作茶叶，第二天已经发酵。为了挽回损失，茶农以当地马尾松干柴进行炭焙烘干，并通过增加一些特殊工序，以最大程度保证茶叶成分。这种为补救过失所制的茶叶即中国最初的红茶制品。这种红茶在国内并不能

▨ 青花瓷茶托上的祁红

撼动绿茶市场，但16世纪末17世纪初（约1604年）由荷兰商人带入欧洲，却随即风靡英国皇室乃至整个欧洲，并掀起流传至今的“下午茶”风尚。

早在鸦片战争之前，英国人已经开始在印度阿萨姆设茶叶种植园尝试种植茶树，经反复试验，于1836年可以生产出少量红茶，并逐年扩大生产。但这时的印度出品红茶无论数量还是质量，还远不能取代中国茶叶。至鸦片战争后的1888年，印度茶叶生产数量及产品质量都有了长足进展，而中国原本主打的武夷岩茶的产量和质量不足以保证市场需求，因而印度出口至英国的红茶产品首次超过了中国。或许，这是安徽祁红“出生”的背景。

Tea and Tea Culture

Chinese tea-drink costom is conservatively estimated to have a history of at least 3,000 years. At the latest, after the advent of the *Tea Classic* by Lu Yu in 780 AD of Tang Dynasty, drinking tea had become an unique habit and cultural phenomenon in China. This costom gradually affected the world. The *Tea Classic* not only recorded the experience of

▨ 远渡重洋到东方进行贸易的船只及港口期待东方贸易品的民众

tea production and tea drinking of China before the Tang Dynasty, but also proposed and summarized a spirit of tea ceremony, namely, clean behavior and thrifty virtue, which made tea along with Chinese calligraphy and paintings a symbol of Chinese culture. Following the *Tea Classic*, Cai Xiang (1012–1067) of the Northern Song Dynasty wrote

▨ 因茶叶交易而带来的巨大的几乎无可逆转的贸易逆差，也被认为是鸦片战争爆发的原因之一

Tea Record during Huangyou period (1049–1053) of the Northern Song Dynasty. The *Tea Record* was divided into two parts to discuss the color, fragrance, taste of tea and cooking methods of tea, such as storing tea, baking tea, grinding tea, sieving tea, boiling and waiting the water, heating the cup, pouring water into the cup as well as tea sets such as utensils for baking tea, tea cage, crushing block, tea clamp, utensils for grinding tea, tea sifter, tea cup, tea spoon, tea bottle, etc. *Tea Record* was the most influential monograph about tea after *Tea Classic*.

Nowadays Chinese mainly makes light tea (green tea, flower tea), but seemingly different ways of drinking tea and corresponding cultural pursuits worldwide today can be found with the original source in the original place of tea, China. Recently, Chinese regard light fragrant green tea as a high-end tea with the basic cooking method of brewing with hot water. But in ancient time, Chinese pursued strong flavor of tea. Chinese tea culture affected the North Korea(D.P.R.), Vietnam, Japan and other neighbors, then spread to Russia, India, and the UK. In the world today, drinking tea becomes a custom and cultural phenomenon in some countries. Russia, the UK, Germany, the US, Turkey, and Argentina can be regarded as important branches of tea culture.

Black Tea and Black Tea Culture

Tea culture can be divided into two factions in accordance with the category of tea leaves: the green tea culture represented by China with the tradition of part-time working while part-time of learning for thousands of years

○ **百年人生，当属始终如一杯温暖茶的巴金（1904—2005），从文化传统到生活习惯，一生与茶息息相关，他被誉为“二十世纪的良心”。**

巴金从小喝茶，日常生活中爱喝沱茶、红茶，还娶了位擅长泡茶的太太萧珊。在巴金身上，上海的“新”和成都的“旧”是人们审视他文学遗产的切入点。1958年的一则巴金与萧珊的书信传达了他们的“茶”生活。是年1月27日，在写给萧珊的信中，巴金说他买了半斤滇红和一斤印度咖啡，问萧珊是否要买祁门红茶等。萧珊回信说，她也买了红茶，是在百货公司买的，有两种价，一种九角，一种五角，每样买了六两，买不买祁门红茶，让巴金做主。一周后，萧珊信中再说，北京有好红茶，不妨再带些。

and oriental poetic tea tasting; the black tea culture (including half black tea) represented by the UK and Russia with the strong flavor adapting to frigid areas and people's pursuit for elegant culture after the industrial revolution.

The tea imported from China maintains its essential fragrance in Russia, but the exotic samovar shows a different subtle tea cultural atmosphere compared with the local one. Subsequent celebrities, such as Tolstoy, Chekhov, Gorky and musician Tchaikovsky, painter Repin and so on, all bred master pieces from this cultural atmosphere.

Different from Russia, the UK represents another important branch of tea culture — Austin refreshing essays and the poems from Byron, Keats and Shelley, which were derived from British afternoon tea.

虽然现代人所用的茶具与古代不一而同，但对于祁门红茶的热爱未尝有丝毫改变

The Born of China's Purely Export-oriented Black Tea

China firstly produced black tea in the Ming Dynasty, and

○ **人间有茶便销魂的郁达夫（1896–1945），在其笔下有很多的人有茶的“歌唱”。**

他端起茶杯，笔下人物也端起茶杯，他放下茶杯，笔下人物也放下茶杯，他喝完茶出门，故事也到了尾声。别人写茶的清淡，他写茶的欲望。别人写泡茶馆的闲适，他写泡茶馆人士的懒散。别人写山中茶的野趣，他写山中茶的逍遥。

there was an accident when taking about black tea. According to legend, in a tea-picking season of one year of middle and late Ming Dynasty, tea leaves picked that day were fermented the next day before they could be made as tea in time due to some reasons in Tongmu area, Chong'an county, Fujian province, Wuyishan region. In order to retrieve the losses, tea farmers used dry firewood of local masson pine to roast the tea leaves and added some special procedures to remain tea component to the maximum extent. This tea born from retrieving loss was the earliest black tea in China. Although this kind of tea could not shake the green tea market in China, the black tea began to prevail in the British royal family and even the whole Europe in the late 16th century and early 17th century (around 1604) and created "afternoon tea" which was passed so far.

篇三

创始在尧渡　创新在共和
——国润茶业的创业

以茶育人，以文化人，在“国润祁红”发展史上占重要位置。司马迁在《史记》中云“王者以民人为天，而民人以食为天”，这是从战争胜负视角论述粮食重要性的。在基本解决温饱问题的情况下，用此论断看茶人茶事有它新的含义。今天，人们在深处优雅宁静的茶室，静静品一杯祁红，感悟茶香“妙境”时，是否知道曾经的前辈们为祁红经历了怎样艰难的发展历程？这里有祁红创始人余干臣（1850—1920）、胡元龙（1836—1924）的贡献榜，更有使祁红在国际上声名鹊起的1915年祁红巴拿马万国博览会捧回的金质奖章。

Passage Three

Founded in Yaodu, Innovated in Republic
—Entrepreneurship of Guorun Tea Corporation

"Educate people with tea while civilize people with litera" holds an important place in the development history of "Guorun Keemun". Sima Qian said "People are everything for a king and food is everything for people" in the **Historical Records**. *It was to emphasize the importance of food from the perspective of victory and defeat of wars. You will find new meanings of people and things about tea by this argument if problem of food and clothing is basically solved. Nowadays, do people know the arduous development process for Keemun that the predecessors had experienced when they enjoy a cup of Keemun in an elegant and peaceful tearoom? These involves contribution from the founder of Keemun, Yu Ganchen (1850 -1920) and Hu Yuanlong (1836-1924) as well as the gold medal obtained on Pan-Panama International Exposition in 1915, which rose Keemun to fame internationally.*

中国茶叶界亟待一个重整旗鼓的转机，而这个转机也不期而至了——祁门红茶适时加入到了外销红茶阵营，并很快继武夷山红茶之后，成为外销的主力。这里，要提及祁红的创始人余干臣（约1850—1920）。余干臣，名昌恺，安徽徽州黟县人士，曾任福州府税课司大使达七年之久。虽然税课司大使仅仅是九品小吏，但因福州是当时全国最大的茶叶出口口岸，故余干臣经常参与到涉及民生的经贸要务。他在任内作为税务官员与以经营红茶为主的公义堂等行帮会首结为朋友，经常前往福建红茶产地，对红茶生产有了明确的了解，深知红茶畅销利厚，为国家重要的经济支柱。1869年，福州茶帮集体抗议洋商压价采购，同时请求政府允许缓缴茶叶税收。亲眼目睹并参与其事的余干臣，虽建言应对外商的蓄意压价，但无奈官卑言轻，在1874年遭罢官变故。

余干臣画像

光绪元年（1875年），“位卑未敢忘忧国”的余干臣自福建回乡，他深知“好山好水出好茶”的道理，处处留意，于是在途中发现家乡（徽州、池州一带）土壤富含有机物，所产茶叶多系楮叶种，叶质肥厚，似乎很适于加工成优质红茶，于是在家乡开始了试制。所制茶叶外形紧细匀整，色泽乌润，当第一批红茶入杯冲泡时，余干臣被那一股从未有过的奇异醇香深深地震撼了，那是一种特有的似花似果似蜜的香气，汤色红艳明亮。余干臣当即让人将红茶捎往福州，如他所料，福州茶帮朋友也被此茶红亮的汤色、沉韵的醇香所震撼，即刻向他建议批量生产。于是，经实地考察，余干臣在产茶区建德县（今东至）尧渡街设红茶庄，开始正式以本地原料仿制工夫红茶。

尧渡街地处尧渡河谷，奇峰叠嶂，古树竹海，乡野散布着茶香弥漫的茶树园。幼读诗书的余干臣至此人杰地灵之地，自然联想到北宋梅尧臣在此任县令时所作《南

▨ 祁门红茶红艳明亮的茶汤

1951 年 10 月 18 日，贵池茶厂全体人员合影

有嘉茗赋》，他当即置办门面，收徒设店，收购鲜叶，并延请宁州师傅舒基立按宁红经验试制红茶，开创性地做起了他所知晓的红茶。与武夷山红茶相比，不啻青出于蓝。

尤为难能可贵的是，余干臣天生具有一般商人所不具备的更为开阔的襟怀：他不是一个保守的传统徽商，他希望他所掌握的红茶技艺不仅令个人受益，更要造福桑梓故里，乃至邦国。他把原本属于独家秘笈的红茶制作方法无偿传授给了更多的人，如陪同他回乡并兼任向导的洪方仁、刘春，并在次年（1876 年）建议祁门人士胡元龙在培桂山房筹建日顺茶厂，而得意门生陈尚好多

▨ 1962 年竣工的润思祁红手工拣厂

年以后学技精湛，也回到自己家乡正冲村自办尚好坊茶号。如同一个播火者，余干臣使这一地区的红茶新品在这一带呈现星火燎原之势，他的回乡之路成为一条探索发现之路，为后人留下了一条幽香甜醇、绝世无双的红丝带。不久，这款在安徽省祁门（徽州辖县）、贵池（今池州市）、东至（池州辖县）、石台（池州辖县）、黟县（徽州辖县），以及江西浮梁（景德镇市辖县）一带均有出产的红茶，而被统称为“祁门红茶”。

不久，祁门红茶穿过蜿蜒曲折的徽商故道，越过重洋，传到英伦三岛，成为英国女王和王室的至爱饮品，由此演绎出风靡欧洲的英式“下午茶”，被誉为“红茶

▨ 1915 年在旧金山召开的巴拿马—太平洋国际博览会旧影

皇后”“茶中英豪”。自此，在与印度红茶、斯里兰卡红茶的竞争中，祁门红茶于民国 4 年（1915 年）获巴拿马万国博览会的金质奖章，由此与印度大吉岭红茶、斯里兰卡乌龙茶并称“世界三大高香红茶”，祁红赢得“茶中英豪”的美称，中国红茶重新赢得声誉和市场，宣告了中国反西方贸易垄断的意图获得成功。在 20 世纪 30 年代，祁红也有少量的“出口转内销”现象。一些留学欧美的知识分子及与外商长期打交道的中国工商界高层人士，随着对欧美文化认识的加深，也开始将英式“下午茶”作为新的时尚的生活方式引进家中。

2014 年秋季某日，李可染之子、著名画家李小可走

展会开幕式

展会中招待中国代表团的宴会

展会金奖奖牌

中国馆入口

进上海克勒门下午茶沙龙，畅谈在北京大雅宝胡同中央美院教职工宿舍与徐悲鸿、齐白石、林风眠、李苦禅、董希文、吴冠中、董永玉等大师“同居”的日子，同来的还有艾青、吴祖光、陆俨少、谢雅柳等人的后代们。茶友的茶文化是印证老一辈文化大家理念的文化，永远无法刻意；做文化，就要以文化之，水到渠成。西风东渐的年月，中国风华绝代的洋派女士们当仁不让地主持着自己的茶聚或沙龙，如林徽因“太太的客厅”声名最为响亮，在北平文化圈内颇具影响力。据后人回忆，20

世纪30年代梁思成的家庭沙龙，延续着每天下午4点半开始喝茶的习惯。林徽因自然是茶会的中心，梁思成则说话不多，偶尔插一两句话，很简洁也很生动诙谐，而林徽因则不管议到什么话题则总可引人入胜，生动活泼。此时，让人回忆起梁、林在西南联大，林徽因对《昆明茶馆》的描述：

这是立体的构画，
描在这里许多样脸，
在顺城脚的茶铺里
隐隐起喧腾声一片。
各种的姿势，生活
刻画着不同方面：
茶座上全坐满了，笑的，
皱眉的，有的抽着旱烟。

今日的安徽国润茶业有限公司（简称“国润茶业”）是国内知名的一家集茶叶种植、加工、品牌运营和国际贸易于一体的茶叶集团企业。从历史沿革上说，国润茶业有限公司及其产品，与余干臣开设尧渡街茶号正是一脉相承。

1950年，中国茶叶公司皖南分公司在贵池县池口村（今池州市池口路33号）筹建大型新式机制茶厂——

▨ 国润祁红的荣誉榜

贵池茶厂。之后，国家对尧渡街余干臣茶号、同春茶号等多家私营祁红老茶号进行了国营化改造，并保存了手工制茶传统工艺。

国润茶业有限公司及其名下的润思祁红，在谱系上确系最正宗的余干臣祁红之嫡传，一直是中国民族工商业界的一面旗帜，但抗战后，尧渡街茶号处于停产状态。

China's tea industry urgently needed to be invigorated and start over again. And a chance showed up—Keemun Black Tea joined the camp of black tea exportation and soon became the main fore of exportation after Wuyishan Black Tea. I'd like to mention Yu Ganchen (1850—1920), the founder of Keemun. Yu Ganchen, alias as Changkai, was born in Qian county, Huizhou, Anhui province. He kept a 7–year tenure of ambassador of Taxation Division of Fuzhou Government. Although the ambassador was only a minor official, Fuzhou was the country's largest tea export port at that time, and consequently, Yu Ganchen often participated in the economic and trade affairs related to people's livelihood. As a tax official, he made friends with leaders of Justice Association and other associations which mainly ran business on black tea. Yu Ganchen often went to producing areas of black tea in Fujian province, so he understood well of black tea production, knowing that black tea were sold well with huge profit. Black tea was an important economic pillar of the country. In 1869, the tea associations in Fuzhou collectively protested foreign merchants for forcing down the price and requested the government to allow deferred tea tax. Yu Ganchen, who witnessed and participated in the protest, insisted on handling deliberate price squeeze by foreign merchants. But he was

出演《茶花女》女主角玛格丽特后尚未卸妆的李叔同（1880–1942）与同学合影

从茶的角度看，他生在茶销区天津，圆寂于茶乡福建，仿佛宿命般与茶有不可割舍的关系，以至于泉州不远的产茶区漳州人林语堂说："李叔同是我们时代最有才华的几位天才之一，也是最奇特的一个人，最遗世而独立的一个人。"1905 年李叔同东渡日本学习音乐、绘画和戏剧，创办戏剧社。为筹集善款，他男扮女装，出演《茶花女》女主角玛格丽特。有人评价，李叔同前半生是走路都掉才华的人，而称弘一法师的后半生（1918 年后），则是精研律宗的慈悲僧人。

dismissed in 1874 due to the abject status.

Before long, the Keemun Black Tea went through the winding routes of Huizhou merchants, went crossed the ocean and finally arrived in the British Isles, becoming the favorite drink of the Queen and the royal family and evolving into popular "afternoon tea" with titles of "Queen of Black Tea" "Hero of Tea". Since then, Keemun Black Tea won the gold medal of the Panama International Exposition in the competition with Indian Black Tea and Sri Lankan Black Tea in 1915. So Keemun Black Tea were known as World's Top Three High-fragrance Black Tea along with the Indian Darjeeling Black Tea and Sri Lankan Oolong Tea. Keemun

巴拿马—太平洋国际博览会中国馆建筑组照

▨ 国润茶厂鸟瞰图

Black Tea won the reputation of "Hero of Tea" and Chinese black tea regained its reputation and market, proclaiming the success of China's anti-western trade monopoly. In the 1930s, "domestic sales of commodities originally" happened to a small amount of Keemun since some intellectuals, who were studying abroad, and senior businessmen, who often dealt with foreign merchants, had begun to introduce British afternoon tea as a fashion of new lifestyle.

Guorun Tea Co., Ltd. of Anhui today (Guorun Tea for short) is a well-known tea group enterprise which integrates tea planting, processing, brand operation and international trade. From the historical perspective, Guorun Tea Co., Ltd.

and its products come down in one continuous line with Yaodujie Teahouse founded by Yu Ganchen.

In 1950, Wannan Branch of China National Tea Corporation set up a large-scale mechanized tea factory with new machine in Chikou village, Guichi county (33 Chikou Road, Chizhou City today). Then nationalization reconstruction for Yu Ganchen Teahoous and Tongchun Teahouse and many other private Keemun old tea houses was carried out, but manual tea craft was preserved.

Guorun Tea Industry Co., Ltd. and affiliated Runsi Keemun are the authentic disciple of Yu Ganchen Keemun according to the pedigree. It is a banner of the Chinese national industrial and commercial circles. However, after the Anti-Japanese War, Yaodujie Teahouse was in a state of suspension.

篇四

青山依旧在　醇香胜往昔
——祁红时光的传承与创新

自祁门红茶问世以来，安徽徽州、池州一带曾有多个种茶基地与厂商。随着时代变换，种茶基地大多保留至今，而工厂则多有凋零。至今仍能保持相当规模者，主要为池州的安徽国润茶业有限公司。从贵池茶厂到今日国润祁红，其历程与变迁深刻地反映了晚清民国到现在中国祁红茶的演变史。从1875年初建的手工作坊到20世纪50年代初国家建厂，“国润祁红”成为“活态遗产”传承经典，有丰富的故事可讲，它是做好文化池州大文章的“活态”点。这座老厂房是“躯壳”，传统工艺是灵魂，留存的是池州耀眼的文化符号，是中国走向世界茶文化的灵气与文化品质。

Passage Four

Present Mellowness Surpasses the Past with —Castle Peak Still In-Inheritance and Innovation of Keemun

Since Keemun Black Tea came out, there had been many tea-planting bases and manufacturers in Huizhou and Chizhou. With the change of the time, most of the tea-planting bases have been preserved so far while most manufacturers have disappeared. Almost only Guorun Tea Co., Ltd. of Anhui survived in Chizhou. From the Guichi Tea Factory to Guorun Keemun today, the historical vicissitude profoundly reflects the evolution history of China's Keemunn Black Tea from the late Qing Dynasty to nowadays. Guorun Keemun became inheritance classic of "live heritage" and there were lots of stories can be told from the original manual workshop in 1875 to the new factory founded by the State in 1950. This is the "live point" for cultural Chizhou. The old factory is the skeleton while traditional crafts are the soul, remaining a glaring cultural symbol. They are aura and cultural quality for China to face the world tea culture.

厂房内堆有尚未入库的茶叶，温润的橙色与厂房的清冷遥相呼应。这些成包的茶叶每天都在来往搬运，是茶厂成为活态遗产的真实写照。

1949年后，中国面临久经战乱而民生凋敝、百废待兴的局面，为了在短时间内恢复生产，在安徽，国家把目光投向了传统的创汇大户——祁门红茶。

然而，形势异常严峻：其一，盛产祁红的安徽徽州、池州一带，因战争重创，茶园凋零，制茶作坊倒闭大半；其二，经1937年之后长达13年的停顿，欧美等传统的茶叶输入大户多半转向其他产茶国；其三，以往的作坊式生产即使恢复，其产量也不足以应付日益扩大的国际市场的需求量。因此，国家要不惜代价恢复祁红的生产与出口，而且要在传统"作坊式"基础上，实施祁红的工业化生产。

于是，国家在安徽徽州与池州两地区筹建三个祁红工厂——祁门茶厂、贵池茶厂和东至茶厂（原贵池茶厂尧渡分厂，于1958年独立）。祁门茶厂在20世纪80年代倒闭，而贵池茶厂和东至茶厂则合并为安徽国润茶业有限公司。

创业伊始——贵池茶厂时代（1950—1976）

中国茶业公司屯溪分公司（后称"皖南分公司"）是筹建新的红茶工厂的执行者。1952年，祁门茶厂、贵

▨ 接受编委会采访的国润茶厂老前辈们，杨焕章（左一）为机械工程师、安徽省劳动模范，袁同昌（左二）为 20 世纪 50 年代制茶车间主任，汪永宽（左三）为 20 世纪 80 年代制茶车间主任

池茶厂和东至茶厂相继成立。祁门茶厂可理解为胡元龙品牌的直系传承，而同在池州地区的贵池茶厂与东至茶厂则为余干臣氏品牌的嫡传。

有关中华人民共和国成立之初贵池茶厂建厂至改革开放更名“国润茶业有限公司”的历程，建厂初期的老一代对其创业艰难仍记忆犹新。他们用口述历史向后人展示了这段历史画卷。

首任厂党委书记作为一位老革命，虽并不精通业务，但热情、开明、爱惜技术人才，全力支持首任厂长徐怀琨开展工作，使得自 1937 年抗战爆发起即销声匿迹的祁红起死回生。现年近百岁的退休老人至今仍记得当初由茶叶种植农户进入工厂成为制茶工人的情景。从茶农到制茶工匠，身份的微妙变化，所激发出的是敬业精神与工匠精神，难忘当年自茶园送茶叶样品至制茶厂，往返需要翻山越岭步行一个昼夜。

老人们几乎异口同声地告诉后人：制作了一辈子的外销红茶，自己却至今没有养成喝红茶的习惯。为什么？红茶是为国家换取外汇的外销产品，尽管是自己亲手制成的，但那是国家财产，自己只有爱惜之情而无权无偿享用。现年92岁高龄的贵池茶厂老职工袁同昌老人回忆道："我1950年3月在东至参加工作，之前是种茶叶的茶农。初期的贵池茶厂条件很艰苦。当时收茶叶都是把大量现金付给农民。那时候没有汽车，山里还有土匪，需要人挑着担子武装押运。工作在山里的茶园人来厂开会，要翻山越岭，一天一夜。刚建厂时年产几千担茶，每年产量稳中有升，20世纪90年代年产量达到30 000担。"

20世纪50年代在我国整体国力积贫积弱的大背景下，贵池茶厂以极其有限的资金投入白手起家，所建厂房仍是精心设计的、高质量施工的，所进口生产设备质量上乘、经久耐用，为后人留下了一份珍贵的工业建筑遗产。当年的贵池茶厂没有辜负人民的希望与重托。1951年4月，贵池茶厂正常投产，很快在国内外市场上重现当年风采，其产品质量甚至是青出于蓝的。祁红于1959年中国"十大名茶"评比会上与南京雨花茶、洞庭碧螺春、黄山毛峰、庐山云雾茶、六安瓜片、君山银针、信阳毛尖、武夷岩茶、安溪铁观音等并列为"中国十大名茶"。至1966年"文革"前夕，毛泽东、周恩来、刘少奇等国家领导人出国访问，祁红都是随行必备的礼品之一。

今天，当人们漫步老厂区，看到老厂房仍可出产手工祁红精品，老库房仍弥漫着茶香，无不赞叹其生产车间、储物仓库之功能完备，而老生产车间的传统工艺之承袭，更令人感叹祁红所蕴含的文化之丰厚。

守业与转型——国润茶业时代（1977—）

作为一家有着光荣历史的大型制茶企业，国润茶业有限公司在这个阶段得到飞跃式的发展。其产品生产由

运行半个世纪之久的制茶机器完全由茶厂工人们自行维护修理

单纯出口转向内外销一体化；由出口供货、贴牌生产转向“以自有品牌为主导，贴牌供货为辅助”的经营模式；由单纯的收购加工转向集茶园管理、初制生产、收购加工、研发创新为一体。贵池茶厂坚守阵地的局面，在为自己赢得了华丽转身的契机时，荣誉也数以十计，主要是：

○ **梁秋实（1903–1987）践行的是文人茶的传统，他留洋回来，办的第一份杂志即《苦茶》。**

梁秋实写过纪念亡妻的小小自传《忆旧篇槐园梦忆——悼念故妻程季舒女士》，文中多处讲茶事。其妻程女士是徽州人，是懂茶之人，她也是梁秋实翻译莎士比亚许多稿子的第一读者。

1983年，祁红获得全国轻工业优质产品评比会金质奖章。

1986年4月，安徽省人民政府授予贵池茶厂“安徽省出口生产重点企业”称号。

1987年9月，祁红荣获布鲁塞尔第26届世界优质食品会金质奖章。

2001年4月，贵池茶厂所出品的润思品牌祁门红茶荣获首届中国（芜湖）国际茶业博览会金质奖。

2003年6月，安徽省茶叶公司贵池茶厂（国有性质）

▨ 茶厂工人在进行手工筛茶工序

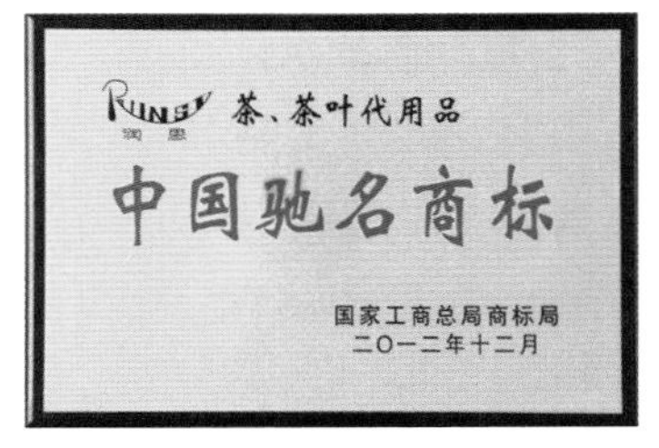

近年国润祁红所赢得的荣誉证书及参与的社会活动（组照）

整体改制为股份制的安徽国润茶业有限公司。

2005 年 10 月，润思祁门红茶荣获首届中国国际茶业博览会金质奖。

2009 年，润思祁红入选 2010 年上海世博会“中国世博十大名茶”，并被指定为联合国馆专用茶。在此次世博会上，时任国家副主席的习近平同志亲临现场品赏润思祁红，联合国秘书长潘基文、芬兰总统哈洛宁、泰国公主诗琳通等多国政要都对润思祁红予以高度评价。

2012 年，“润思”商标被国家工商总局认定为“中国驰名商标”。

2015 年 6 月，润思牌祁红再次夺得北京国际茶业展红茶特别金奖。

2016 年“润思红茶”被国家质检总局认定为“中华人民共和国生态原产地保护产品”。

相比这些耄耋老者，现任董事长殷天霁先生是“文革”结束后大学毕业来厂任技术员的“80 年代新一辈”，同样对茶厂倾注了无条件的爱。他向友人回忆那段往事：

“从 1994 年开始，因为增值税改革，对企业来说是大的挑战，从国内来说国内绿茶市场兴起，对红茶市场有一定冲击……欧盟出台了高标准的进口标准，对中国茶叶出口是重大考验，其标准比国内标准高百倍。当时中国红茶主要出口欧盟，那一年很多厂家面临严峻考验，2000 年以后，祁红厂家雪上加霜，更加艰难。为此，贵池茶厂必须做出改革。幸运的是，国润一直航行在入欧的海上丝绸之路中。”

“……仅以老毛茶仓库为例，它的四壁、地板与屋顶无处不浸满了几十年茶香，积淀着几代人的辛勤劳作，不能忍心在我们这一代人手中消失！”

“我能够回到这个厂子里，其实当时首先是有勇气，其次要有自信。当时市里同行很多做绿茶，说红茶赚不到钱，我认为他们对祁红缺乏了解，我知道祁红在国外的影响力，要看到它的潜在价值。我相信，世界三大高香茶不可能一

▨ 从手工拣厂外壁的结构柱上可以看出，原本一层的平房因茶厂发展需要而进行了加建

直被埋没，这么好的东西在我们手上毁掉了，就是犯罪！我承接下了银行的债务，这在当时改制的风潮下是极其罕见的，但这也种下了一颗诚信的种子，使我们进入良性运转；相信国润祁红一定是好的，要有公心；在改制之前，大部

分的祁红企业依赖于出口外贸，我们开始用润思品牌做国内品牌，不局限于出口，市场变动承受能力变强；2003 年改制，国营贵池茶厂改称为“国润茶业有限公司”，其实是民营控股的有限责任公司。我们不是作为改制的既得利益者，而是致力于祁红的振兴和发展，是与全体职工同甘共苦，这是我们的自律。”

“比较荣幸的是，改制过后我们的一些制茶方面的技术顾问都留了下来。这些人是难得的财富。再有对老厂房的保护，我们直觉就应该保持原貌，也就是修旧如旧。那时候从老国企过来的职工管理习惯还是好的，机器保养、厂房清理，好的习惯无意中延长了老建筑、老设备的寿命。我们的感悟是：首先，老的祁红厂和工艺保留下来了；其次，我们实现了转型，坚持外销，并转型内销。在祁红行业，我厂是第一个以品牌导入做内销的。2010 年我们亮相上海世博会，唤醒了祁红厂区，振奋了祁红精神。”

“2017 年我们又入选了中国 20 世纪建筑遗产名录，这说明祁红作为特殊的符号，是无可替代的。”

润思祁红何以历久弥新

选料精良。润思系列产品的祁红，原材料的茶叶出自地处北纬 29°~30°、东经 117°~118° 之间的黄山、

入选第二批中国 20 世纪建筑遗产名录后，厂区制作了带有标识性的导视牌

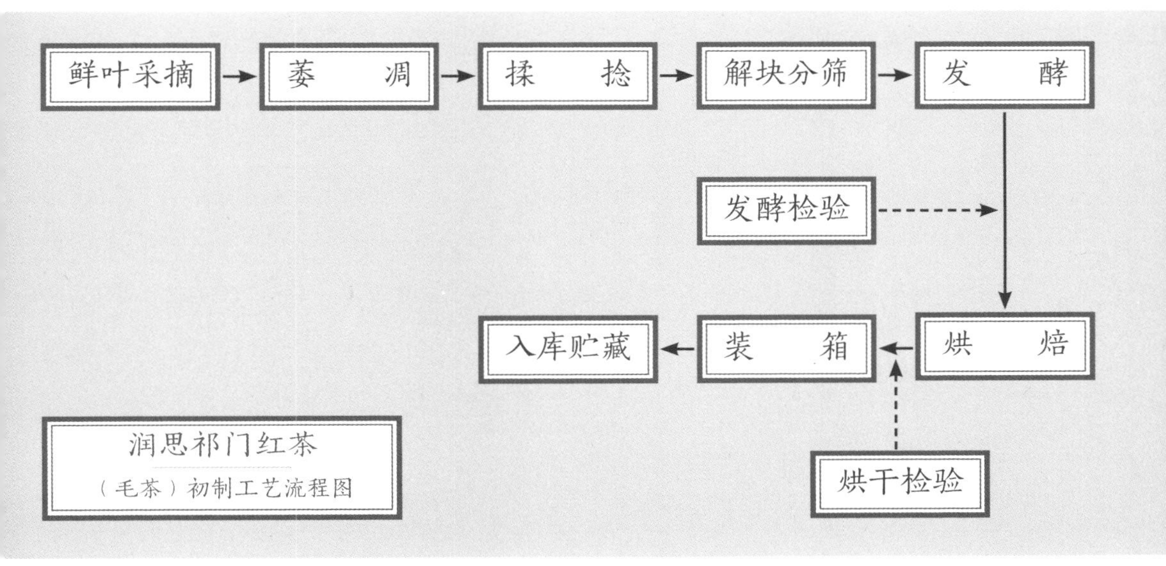

润思祁门红茶
（毛茶）初制工艺流程图

九华山一带的黄金产茶区。此地为山地，适宜茶树生长并出品优质茶叶的地带为海拔 600 m 左右的坡地，年平均降雨量为 1 650~2 000 mm，土质为石砾含量、透水透气程度均良好且肥力较高的红黄壤。经过几十年的考察和筛选，先后建立了同春茶园、黄山乌石茶园和古溪茶园等三个茶叶种植基地。1984 年，全国茶树良种审定委员会审定祁红种植基地的槠叶种为“全国茶树良种”。

独特的加工工艺。润思祁门红茶生产工艺分为粗制工艺和精制工艺两种。

润思祁门红茶的精制工艺，复杂而精细。一般采用单级付制、多级收回的方式进行，毛茶付制毛火烘干后，用联装作业机取料，然后按照本身路、长身路、圆身路、轻身路四路单机制作。各路头尾茶单独处理。整个作业流程大致包括投料付制、毛火烘干、切分取料（毛筛、抖筛、切轧）、筛分（园筛、紧门、撩筛）、风选、拣剔（机拣、色选）、清风补火、拼和匀堆、成品装箱等十多道工序。

这个红毛茶制成后的精制阶段，在工艺上要求自然又比前一阶段更为精细严苛。首先，要将长短粗细、轻重曲直不一的毛茶经过筛分、切轧取料后整饰外形，风选、拣剔后去除杂质，制成嫩度均匀、长短整齐、大小一致的孔号茶半成品，最后按照商品茶的品质要求，将

润思祁红茶园一隅

众多孔号茶按照一定比例进行拼和匀堆，成为成品茶。同时为了提高干度，保持品质，便于贮藏和挥发茶香，还要进行补火等烘干处理，最后，通过装箱审评检验后，才成为形质兼优的润思祁门红茶商品

经过初制、精制两个工艺加工流程，即产生出多种多样的适应不同消费群体的红茶系列产品。其最上等的精品，可作“国礼”，而由圆身路等产生出的一般性产品，则是价廉物美的大众日常消费产品。

茶园工人在采摘鲜叶

多元一体的“国润祁红”的发展策略

1. 健体功能的作用传播

传统中医认为：祁红甘温，可养人体阳气、生热暖胃。而民间早已将祁红作为暖胃助消化的良药。在贵池茶厂——国润茶业的老职工中，尽管大多没有饮红茶的习惯，但身体出现消化不良状况时，厂医务室会将祁红作为医疗处方。同时他们不断邀请中外食品营养学、医药学的专业团队对所有润思系列产品作严苛的检验分析。

早在 1995 年，经与国际机构联络，在英、美、加拿大等国开展了“祁红对人体健康的作用”的研究。国外有研究报告指出：祁红的疗效虽然无法使病人的血液流通恢复正常，但有助于改善血管畅通状况。祁红中的茶黄素和茶红素等氧化缩合物具有抗氧化、抗自由基的作用；祁红中的茶多酚类化合物可以降低血压中的胆固醇，明显改善血液中高密度蛋白与低密度脂肪白的比值；

祁红中的咖啡碱能舒张血管、加快呼吸、降低血脂；祁红除含多种水溶性维生素外，还富含微量元素钾，经冲泡，70% 的钾可溶于茶水中，增强心脏血液循环，并能减少钙在体内的消耗。因此，祁红对防治中风、心脏病和脑血管病等是具有相当的功效的。此报告谨慎做出结论：每天喝 5 杯红茶的人，心脏病患者血管舒展度可从 6% 增加至 10%，而常人的血管舒展度能增加到 13% 左右；脑中风的发病危险比不喝红茶的人低 69%。

2. 祁红文化应传播

中国茶叶历史悠久、种类繁多，有传统名茶和历史名茶之分，所以中国的“十大名茶”在过去也有多种说法。

1915 年巴拿马万国博览会将碧螺春、信阳毛尖、西湖龙井、君山银针、黄山毛峰、武夷岩茶、祁门红茶、都匀毛尖、铁观音、六安瓜片列为中国十大名茶。

1959 年中国“十大名茶”评比会将南京雨花茶、洞庭碧螺春、黄山毛峰、庐山云雾茶、六安瓜片、君山银针、信阳毛尖、武夷岩茶、安溪铁观音、祁门红茶列为中国十大名茶。

2002 年《香港文汇报》将西湖龙井、江苏碧螺春、安徽毛峰、湖南君山银针、信阳毛尖、安徽祁门红、安徽瓜片、都匀毛尖、武夷岩茶、福建铁观音列为中国十大名茶。

2009 年，祁红入选 2010 年上海世博会“中国世博十大名茶”。

▨ 手工筛茶的筛网架上，“中茶贵池直属”“公元一九五三年四月”的字迹隐约可见

3. 润思祁红是文化池州“新名片”

讲好祁红故事，是做好文化池州“名片”的目标。历史上西方的《荷马史诗》或中国的《史记》能够传承至今，就是用故事承载了文化的方方面面。将生动的“茶”文化融进故事中，会让品茶者百听不厌。从文化池州的“茶”故事看，可讲的内容很丰富，如茶与养生、茶与美容、茶与历史、茶与礼仪、茶与中外古今名流、茶与建筑文化、茶与美术音乐、茶与文化发展的“解码”等。茶韵之所以美好，是因为她深藏着一切；茶香之所以美好，是因为香茶回味难忘。

将已入选中国20世纪建筑遗产名录的祁红茶厂所产茶之高香、高贵、高品质的特质，尽情表现在文化池州建设上是一项艰巨的任务。那么祁红的清新典雅、从历史走来的风范感染今人的妙绝，就需要我们从找寻祁门茶厂老厂区的消失，即令毗邻的国润茶业也为之扼腕惋惜的历程入手。发现老厂区，也是发掘联系中外古今的祁红文化：润思祁红是有生命有灵魂的文化产品，而地处池州市的老厂房与东至县尧渡街老茶号等，就是饮水思源的润思祁红之源头，必须加以珍惜。品润思祁红能令公众联想到一家沿袭近70年却仍在运作的老厂房，是怎样写意的“乡愁”故事。

由此，从发展中国、面向世界的红茶文化入手，为什么我们不能设想拥有中国倡导主持的世界红茶博物馆联盟？为什么我们不能借“一带一路”走好中国祁红的传承与发展之路？为什么不能找回中国的祁红令世界瞩目的目光？我们有理由说：祁红那由心底释放出的醇香，会让每位品到的人皆为之陶醉并喜爱上它。

润思祁门红茶入选“中国世博十大名茶”

▨ 润思祁红在“世界和谐茶会”活动中接受来宾品鉴

▨ 专家在拣厂一侧的展示室中品鉴祁红茶

After 1949, China were trapped in a situation where people lived in destitution and thousands of things needed to be done after the war. In order to recover production within a short period, the country turned its attention to the traditional foreign exchange earner—Kemmun Black Tea in Anhui province.

However, the situation was extremely serious: Firstly, Huizhou and Chizhou areas of Anhui province, which abound with Keemun Black Tea, suffered a serve damage in the war. Tea-planting bases withered and more than half of the tea-making workshops collapsed. Secondly, Europe and America turns to other tea-produced countries after China's 13 years of stagnate after 1937. Thirdly, even if the previous workshops got recovered, their poor output would not meet the expanding demand from international market. Therefore,

China would restore the production and export of Keemun Black Tea at all costs and implement industrial production based on the traditional “workshops”.

As a result, China established three Keemun Black Tea factories in Huizhou and Chizhou in Anhui Province—Keemun Tea Factory, Guichi Tea Factory and Dongzhi Tea Factory. The Keemun Tea Factory collapsed in the 1980s while the Guichi Tea Factory and the Dongzhi Tea Factory merged into Guorun Tea Industry Co., Ltd. of Anhui.

In 1983, Keemun won the gold medal of Quality Product of National Light Industry Committee.

In April 1986, the People's Government of Anhui Province awarded Guichi Tea Factory the title of “Key Export Enterprise of Anhui Province”.

In September 1987, Keemun won the gold medal of the 26th World Quality Food Association of Brussels.

In April 2001, Keemun of Runsi brand produced by Guichi Tea Factory won the gold medal of the First China (Wuhu) International Tea Industry Expo.

In June 2003, Guichi Tea Factory of Anhui Tea Corporation (state-owned) was overall restructured into the joint-stock Guorun Tea Co., Ltd. of Anhui.

In October 2005, Runsi Keemun Black Tea won the gold medal of the First China International Tea Industry Expo.

▨ 三位国润茶厂老前辈接受编委会采访

In 2009, Runsi Keemun Black Tea was selected as one of the "Top Ten Famous Teas of China World Expo" at the 2010 Shanghai World Expo and was designated as the exclusive tea for the UN Pavilion. At the World Expo, Comrade Xi Jinping, the Vice President of the State then, personally visited the Expo and appraised the Runsi Keemun Black Tea. UN Secretary-General Ban Ki-moon, Finnish President Halonen, Thai Princess Sirindhorn and many other foreign dignitaries all highly praised Runsi Keemun Black Tea.

In 2012, the trademark of "Runsi" was recognized as "China Famous Brand" by the Country State Administration for Industry and Commerce.

In June 2015, Runsi Keemun Black Tea once again won the special gold medal for black tea of Beijing International

Tea Exhibition.

In 2016, "Runsi Black Tea" was recognized by the General Administration of Quality Supervision as "eco-origin protection product of the People's Republic of China".

Compared with these elders, Mr. Yin Tianji, the current chairman, was the man of "new generation of the 1980s" who graduated from the university after the "Cultural Revolution" and worked as a technician in the factory, pouring his unconditional love into the tea factory.

2017 年 10 月，92 岁的老制茶车间主任袁同昌回到检验楼中，为大家演示当年的手工验茶的方法

祁门红茶厂房烟囱剪影

Pluralistic Development Strategy of "Guorun Keemun"

TCM (traditional Chinese medicine) has long believed that: Keemun is lukewarm, which can raise body's yang–qi and warm the stomach with heat. Folks have used Keemun as a good medicine to warm stomach and aid digestion. Among the old employees of the Guichi Tea Factory—Guorun Tea Corporation, although most of them do not have the habit of consuming black tea, the factory clinic will use Keemun as a medical prescription when the employees get dyspepsia. Meanwhile, they constantly invite Chinese and foreign food nutrition and pharmaceutical professional teams to conduct rigorous examination and analysis on all Runsi products.

▨ 茶工们拣厂中拣茶

Runsi Keemun is the "new name card" of cultural Chizhou.

The goal of making a good name card of Chizhou is to tell the stories of Keemun well. Historically, the reason why the *Homer's Epic* in the West or the *Historical Records* in China could be passed down until now, because the books carries all aspects of culture via stories instead of dull empty talk. People will never get bored of listening to tea if the tea is incorporated with stories. From the tea story for the

▨ 检验楼中的茶叶样本和检验工具

cultural Chizhou, there are many things can be told, such as tea and health-preservation, tea and beauty, tea and history, tea and etiquette, tea and celebrities from China and foreign countries, tea and architectural culture, tea and arts as well as music, the decoding of tea and cultural development and so on. The reason why tea rhyme is so wonderful is that it deeply embraces everything. The reason why tea fragrance is wonderful is that it is unforgettable despite its short aroma.

It is a tremendous challenge to fully express features of high fragrance, nobleness and high quality of Keemun Black Tea, which has been selected into the China's 20th Century Architectural Heritage Projects, on brand construction of cultural Chizhou. We should begin from the disappearance of old factories of Keemun Black Tea, which is also a progress regretted by adjoining Guorun Tea Corporation, to find out why the elegance and historical manner do inspire people today. Discovering the old factory is to explore the culture of Keemun Black Tea that connects the ancient and modern China and foreign countries: Runsi Keemun is a product with life and soul, old factories in Chizhou and Yaodu street Teahouse in Dongzhi county are the source of Runsi Keemun, which should be cherished. Consuming Keemun Black Tea can lead people to think of old factories with a history of 70 years. This is the story of "remember nostalgia" that should be well written cultural Chizhou, even China.

Therefore, why can't we imagine to set up an Alliance of Global Black Tea Museums hosted by China from the black tea culture which develops China and faces the world? Why can't we take the road of inheritance and development of China's Keemun by "the Belt and Road"? Why can't we retrieve that Keemun which draws attention of the world? We have reasons to claim that Keemun can make fragrance risen from the bottom of people's mind and every single person who drinks it will be intoxicated and fall in love with it.

编后 Afterword

“国润祁红”是发生并创造“奇迹”之地

“Guorun Keemun” Is a Place Where “Miracle” Is Created and Occurs

2018 年 3 月 15 日，在第二批中国 20 世纪建筑遗产地——故宫宝蕴楼——召开“安徽国润茶业有限公司创意设计调研报告”研讨会，该会议由中国文物学会 20 世纪建筑遗产委员会与池州市人民政府合办举行。恰如单霁翔院长所说，其成果针对中国 20 世纪建筑遗产项目优秀个案的研究与推广会，它用祁红全遗产向世人展示物质与非物质兼得一身的内涵与精髓。池州市雍成瀚市长在赞赏国润祁红老厂房被中国文博建筑专家发现的独到慧眼时，希望通过专家建言使它“活起来”，造福池州人民，也为中华茶文化走向世界有所贡献。

2018 年 5 月 19 日，中国建筑学会在泉州举办 2018 年学术年会，特在泉州威远楼广场举办“中国 20 世纪建筑遗产名录展”，国润祁红旧厂房与有百年历史的中国工业遗产首都钢铁公司并列呈现，表现了国润祁红遗产的文化特质，吸引了数以万计市民驻足的目光。2018 年 5 月 20 日，在泉州海上交通博物馆（已故建筑学家杨洪勋设计）举办的“新中国 20 世纪建筑遗产的人和事学术研讨会”上，笔者做了《中国 20 世纪建筑遗产是亟待保护利用的新蓝海——关于建筑遗产事件与人的敬畏与回望》发言，讲述了国润祁红厂入选中国 20 世纪建筑遗产的理由。

中国茶文化的源远流长取决于文人的参与，其生活美学对茶文化的作用是决定性的。器往往不嫌其美，物往往不厌其精，行往往不讳其细。现在思考，本人之所以沉溺于祁红茶的生产场景建筑中，有以心逐物之感，是因为从一开始便体味到

▨ 2017 年 8 月，专家团队随修龙理事长首次造访国润祁红

它不平凡的文化内涵。我是在中国建筑学会理事长修龙引导下，于 2017 年 8 月 24 日—8 月 26 日在第一次到池州，便与同行专家们发现并喜欢上国润祁红旧厂房的。其感知至少有三点。

第一步，发现与认定之旅。从 2017 年 8 月 24 日第一次走进国润祁红厂，由对朴素之美的工业遗存的“活态”感悟，联想到中国20世纪建筑遗产的评定标准，体会着它传承至今的珍遗，更从殷天霁厂长的谈吐中多了一份感动和自信。2017 年 10 月初在池州市人民政府支持下，考察组的调研、访谈、摄影与一次次研讨不断展开。

第二步，发现与评估。2017 年 10 月初，中国文物学会 20 世纪建筑遗产委员会便组织了专家团队，并没有简单以国润祁红旧厂房入选“第二批中国 20 世纪建筑遗产项目”为目标，深度在于用此“引爆”文化池州的“城市文化工程”。一方面要从学理上充分论证“国润祁红”的厂房建筑体现了新中国初创时是怎样的风采，这个面积与产值都不大的“茶厂”何以成为中国 20 世纪建筑工业遗产的特殊名片；另一方面，利用品质名扬天下的祁红茶，串接起中国正实施的“一带一路”发展之策，创出有社会文化“大价值”的祁红格调、情感、心态与生活方式。曾四次出任英国首相的格拉德斯通（1809—1898）有句名言：“如果你发冷，茶会使你温暖；如果你发热，茶会使你凉快；如果你抑郁，茶会使你欢快；如果你激动，茶会使你平静。”

第三步，设计考察与祁红品牌传播。

▨ 2018年3月，考察组在国润祁红老茶厂合影

▨ 2018年5月，在泉州威远楼广场上参展“中国20世纪建筑遗产”的国润茶叶祁门红茶旧厂房

继2018年元月向池州市人民政府提出12条《国润祁红贵池茶厂老厂区创意设计规划要点》后，我们于2018年3月3日—5日组织全国设计大师、著名规划建筑文博专家再度考察“国润祁红”，旨在为3月中旬在故宫宝蕴楼召开研讨会做好准备。

为传播祁红“故事”，2018年4月下旬我们再赴池州祁红，与市文广新局、规划局领导共议“国润祁红”事项，当置身于霄坑茶田，感悟到唯有绿水青山，才有生态好茶；唯有以禅入茶，才有文化立基茶禅的道之境界。这是一本试图要讲好“国润祁红”故事与品质的书。为此，需要再介绍一个事实：

据报载，2015年10月20日下午17时，习近平夫妇访英期间与查尔斯王子在克劳伦斯宫一同享用著名的英式下午茶，这茶一般选用祁门红茶再加工。对中国人说来，不论是开门七件事的“柴米油盐酱醋茶”，还是文人雅士的“琴棋书画诗酒茶”，中国文化与生活绝离不开茶，英国人喝茶以红茶居多，下午茶则是另一种文化。2015年10月20日，在英国白金汉宫国宴上，习主席致辞中说：“中国的茶叶为英国人的生活增添了诸多雅趣，英国人别具匠心地将其调制成英式红茶。中英文明交流互鉴为人类发展做出了巨大贡献。”习主席在英国传递的祁红“故事”，说明祁门红茶确有理由带给中国与世界一道靓丽的风景。

为了国润祁红旧厂房入选中国20世

纪建筑遗产项目名录，为了理解文化池州的“茶人茶事”感人故事，《中国建筑文化遗产》编辑部专家团队至少八次造访池州贵池工厂，在念往事中，话祁红，意在为祁红故事掀开新篇。如今贵池厂区已能看到有导向标识的“国润祁红”的展示平台，说明祁门红茶的传播已走上日程。早在2017年初，外交部部长王毅在品“国润祁红”茶后，就评价其为“镶着金边的女王”，这是多么到位的中国茶文化的宣传语呀！为今人建构起一种更丰富的茶生活体验：走千年百年茶路、住茶农的“客房”、吃茶的餐饮、品茶的滋味、听茶的故事、讲茶的文化、享茶的禅韵等，都希望用茶带来健康和愉悦……

我特别想说的是，与其说是“国润祁红”旧厂房的精湛打动了我，倒不如说是殷厂长独有的执着与坚守感染着我。如果站在改革40年的“节点”上看，贵池厂留下的是匆匆的开放步伐，是“国润祁红”人不屈的发展足迹；如果站在中华人民共和国70年的“史实”上想，老房屋留下的亦建筑、亦工艺、亦什物的传承，就是一组新中国工业文化的“绝照”。当我们几次派出摄影师为那弥漫着茶香的“仓库”拍照时，从那一幅幅泛红色的照片中，我们看到了祁门红茶的“颜色”，更体会到殷厂长为什么拒绝了所有开发商，坚守到“文化池州”开启建设这一天。因为池州有数十万茶农、茶工的“乡愁”，透视出全民族对中华茶文化美味的尊崇与自觉。

“国润祁红”何以能发生并创造着“奇迹”，它给世间仅仅是以茶学师、以茶悟心、以茶缘友吗？仅仅是贵池老厂房“活态”使用中透析出的历史记忆与当代启示？我以为对遗产传承与创意皆为使命的当下，它的奇迹是要在旧传承中建设新标杆，在告慰中国茶文化的同时拓展到东方文化与文化世界。

我坚信这里会出现中国乃至世界的“祁红”新品牌，因为在“一带一路”精神下，它必将成为聚拢全球三大高香茶的世界茶文化博览园。“国润祁红”的文化池州模式有它发生的逻辑，更有它迈向成功的理由。所有这些均成为我们团队着力研究、品读、传播推荐它的动力。

金磊

中国文物学会20世纪建筑遗产委员会

副会长、秘书长

2018年7月

On March 15, 2018, conference of Research Report of Creative Design with Guorun Keemun Tea Co.,Ltd. of Anhui was held in Baoyun Building, which was one of the Second Batch of China's 20th Century Architectural Heritage. It was co-held by the 20th Century Architectural Heritage Committee of Chinese Cultural Relics Society and the People's Government of Chizhou. As Dean Shan Jixiang said, the outcome of the conference was to study and promote the excellent case of China's 20th Century Architectural Heritage Project. It showed the connotation and essence of the combination of both material and non-material by "full heritage". The mayor of People's Government of Chizhou appreciated the unique perspective of Chinese cultural and architectural experts for discovering the old factory of Guorun Keemun. What's more, he hoped that the old factory could be "alive" by experts' recommendations. Hopefully, it can benefit the people in Chizhou and also make a contribution to Chinese tea culture at the same time.

Why can Guorun Keemun create the"miracle"? Is what it brings to the world just only helping people get educated, enlightened and obtain friendship with tea? Is that just the historical memory and contemporary enlightenment reflected in the using process of "live" old factory of Guichi? Nowadays, its miracle needs to establish a new guide post in the old heritage and expand oriental and world culture while comforting Chinese tea culture with missions of inheriting heritages and innovation.

Jin Lei
Vice President & Secretary General of 20th Century Architectural Heritage Committee of Chinese Society of Cultural Relics

《悠远的祁红——文化池州的“茶”故事》编委会

主编单位　中国文物学会20世纪建筑遗产委员会　池州市文广新局
池州市城乡规划局　安徽国润茶业有限公司

学术顾问　吴良镛　谢辰生　关肇邺　李道增　傅熹年　彭一刚　陈志华　张锦秋　程泰宁
何镜堂　郑时龄　费　麟　刘景樑　王小东　王瑞珠　黄星元

名誉主编　单霁翔　修　龙　马国馨

编委会主任　单霁翔

编委会副主任　雍成瀚

主编　金　磊　贾　瑄

策划　金　磊

编委　王建国　徐全胜　王时伟　付清远　孙宗列　孙兆杰　朱　颖　伍　江　刘伯英
刘克成　刘若梅　刘　谞　庄惟敏　邵韦平　邱　跃　何智亚　张立方　张　宇
张　兵　张　杰　张　松　张爱林　李秉奇　杨　瑛　陈　薇　陈　雳　陈　雄
季也清　赵元超　徐　锋　郭卫兵　殷力欣　周　岚　周　恺　孟建民　金卫钧
常　青　崔　愷　梅洪元　奚江琳　路　红　韩振平　江　心　吴　飞　胡军保
林　芳　罗祖奇

执行主编　殷力欣　李　沉

撰文　殷力欣　金　磊　张明汉　汪松柏　汪贵帆　苗　淼　崔　勇　林　娜　陈　雳
朱有恒　（部分文稿由安徽国润茶业有限公司提供）

执行编辑　朱有恒　吴　衡　董晨曦　李海霞　李　玮　郭　颖　王　展　金维忻

美术编辑　朱有恒　谷英卉

图片提供　安徽国润茶业有限公司　部分照片来自网络

建筑摄影　中国建筑学会建筑师分会摄影团队：李　沉　朱有恒　金　磊　殷力欣　等

图书在版编目（CIP）数据

悠远的祁红：文化池州的“茶”故事 /《中国建筑文化遗产》编辑部编. -- 天津：天津大学出版社，2018.9
（中国20世纪建筑遗产项目·文化系列）
ISBN 978-7-5618-6249-0

Ⅰ. ①悠… Ⅱ. ①中… Ⅲ. ①红茶—介绍—池州
Ⅳ. ① TS272.5

中国版本图书馆CIP数据核字（2018）第216071号

Youyuan de Qihong: Wenhua Chizhou de “Cha” Gushi

策划编辑：金　磊　韩振平
责任编辑：郭　颖
装帧设计：朱有恒　谷英卉

出版发行　天津大学出版社
地　　址　天津市卫津路92号天津大学内（邮编：300072）
电　　话　发行部：022-27403647
网　　址　publish.tju.edu.cn
印　　刷　北京华联印刷有限公司
经　　销　全国各地新华书店
开　　本　165mm × 230mm
印　　张　8
字　　数　103千
版　　次　2018年9月第1版
印　　次　2018年9月第1次
定　　价　49.00元